Seabed Minerals Series

Volume 2

Analysis of Exploration and Mining Technology for Manganese Nodules

Seabed Minerals Series
Volume 2

Analysis of Exploration and Mining Technology for Manganese Nodules

**United Nations
Ocean Economics and Technology Branch**

Published in co-operation with the United Nations
by Graham & Trotman Limited

Published in 1984 by
Graham & Trotman Limited
Sterling House
66 Wilton Road
London SW1V 1DE
in co-operation with the United Nations

© The United Nations, 1984

Softcover reprint of the hardcover 1st edition 1984

British Library Cataloguing in Publication Data

Analysis of exploration and mining technology
for manganese nodules.—(Seabed minerals
series; v. 2)
1. Manganese mines and mining, Submarine
2. Manganese nodules
I. United Nations. *Ocean Economics and
Technology Branch*   II. Series
622'.34629'09162   TN490.M3

ISBN-13:978-94-010-8980-7   e-ISBN-13:978-94-009-5622-3
DOI: 10.1007/978-94-009-5622-3

Typeset in Great Britain by Acorn Bookwork, Salisbury, Wiltshire

Contents

Preface to the Series

The deep seabed is one of the newest and potentially most rewarding frontiers that has challenged mankind in its quest for knowledge and material achievement. Resources of the deep seabed promise to make an enormous contribution to the world's resource base if their potential is realized. At the present time, the resources of the deep seabed of immediate interest are in the form of manganese nodules which lie on the surface of the ocean floor and contain numerous metals — copper, nickel, cobalt, manganese, molybdenum, vanadium and titanium.

In addition to the potential for increasing the world's resource base, these minerals are particularly intriguing because they lie beyond the limits of national jurisdiction – they belong to no nation. According to a resolution adopted by the General Assembly of the United Nations, the area of the seabed and the resources therein are the common heritage of mankind.

While, for many years, manganese nodules have been the domain of marine geologists, chemists, and lawyers, interest is currently being shown by a broader audience — corporate and public entrepreneurs, policy-makers, managers, investors, engineers and technicians, as well as the international community of persons interested in the mineral and ocean industries, strategic minerals, and global development. The past two decades have witnessed the development of technology with a view to mining manganese nodules from a depth of nearly 15,000 feet and to extract the economically important metals from them. It has also witnessed the development, in the Third United Nations Conference on the Law of the Sea, of the legal and institutional framework in which the exploitation of the nodules will take place — who will exploit these resources, how this will be done, and who will benefit from this exploitation. At this moment, a watershed appears to have been reached when decisions about application of technological prototypes of commercial operations are being taken and when the negotiations in the Law of the Sea

Conference culminated in the adoption of the United Nations Convention on the Law of the Sea. Thus, it seems to be an appropriate time to take stock, to assimilate widely diverse and dispersed information and analyses, and to make preparations for a future which envisages the resources of this newest frontier being utilized for the benefit of mankind as a whole.

It is in this context that the United Nations has undertaken the task of preparing a series of nine volumes examining different aspects of the development of manganese nodule resources.

The series starts with a discussion of how much of these resources exist in the world ocean. The quality and the quantity of the data collected through scientific expeditions and prospecting and exploration cruises are examined, the methods of estimating resources and reserves and their distribution on the basis of these data are explained, and a review of the currently available estimates is presented.

Since new technology had to be developed and tested for exploring, mining and processing manganese nodules, the next two volumes in the series describe and evaluate these technologies in terms of their efficiency and reliability.

Mine development is preceded by the delineation of a mine site. The fourth volume discusses the criteria used to designate a mine site. A potential mine-site is defined as satisfying certain economic, geological and technological criteria; the size of the area may vary according to these criteria. This volume also develops and applies a simulation model which calculates the variations in the size of the mine-site resulting from the variations in these criteria. Finally, based on available data and with the application of the simulation model, this volume attempts to identify areas meeting given sets of criteria.

The location of a processing plant for manganese nodules depends more on the availability and costs of complementary inputs, e.g. energy, chemicals, than on the traditional considerations of proximity to the mine or to the market. The criteria for selecting a site for a processing plant are examined in the fifth volume and a few likely sites of first generation processing plants are assessed according to these criteria.

The economics of a manganese nodule mining project is the crucial factor in the development of these resources. Currently, the economics can at best be based on order-of-magnitude estimates. These estimates are carefully examined and a financial profile of a potential seabed mining company is constructed in the sixth volume.

The seventh volume attempts to compare the economics of a manganese nodule mining project with alternative mining projects, e.g. nickel laterites.

The geological, technological and economic considerations for resource development can be translated to practice only within a defined regulatory framework. The eighth volume reviews the regulatory framework for the exploration and exploitation of seabed resources that is being devised by the international community. Finally, private companies, joined in several consortia, and public entities have already initiated and are carrying out work on mine development; the last volume of the series presents a summary of their activities and the status of their work.

The series is being prepared by the Ocean Economics and Technology Branch of the United Nations Department of International Economic and Social Affairs together with leading external experts. It draws and builds upon the information, data and expertise that the United Nations has been developing for over a decade concomitant with its work in the Law of the Sea Conference, and currently, in the Preparatory Commission for the International Sea-Bed Authority and for the International Tribunal for the Law of the Sea. These capabilities have been utilized in, and at the same time strengthened by, the development and maintenance at United Nations Headquarters of a data base on manganese nodules and a literature data base on marine minerals as well as the convening of a Group of Experts Meeting, the proceedings of which were published as *Manganese Nodules: Dimensions and Perspectives* (Dordrecht: Reidel, 1979).

The efforts of the United Nations will succeed if the series can contribute to wider understanding of seabed resources and the problems and opportunities involved in their development, and can encourage constructive ideas and actions.

Acknowledgements

We gratefully acknowledge the contributions of Science Applications, Inc., in particular P. Grote and W. Coleman, and of E. Langey in the preparation of this volume. We are particularly appreciative of the kind permission given to us by the Bureau of Mines, United States Department of the Interior for using some materials prepared for the Bureau. The contribution of E. Abrams, a United Nations intern, in drafting a part of Chapter 4, is appreciated. Some parts of the manuscript draw heavily on informal papers by L. Antrim and D. W. Pasho, and we are grateful to them. We thank L. Antrim, S. Boshkov, J. S. Chung, D. W. Pasho and W. D. Siapno for reviewing the manuscript and offering many helpful comments and suggestions.

Chapter 1

Introduction

Oceans have been the subject of scientific inquiry for hundreds of years, but significant study of mineral occurrences on the deep ocean floor has only begun to take place. Man's present knowledge of the ocean floor had to await the development of sophisticated research equipment capable of probing the ocean to great depths. This began in the 1940's and since then the accelerated pace of ocean research has generated a large amount of data on the ocean environment — mostly through the work of academic and governmental scientific organizations around the globe. These new scientific disclosures confirmed the wide-spread occurrence of metal-bearing lumps on the deep ocean floor that hold great promise as an important new source of raw material.

Encouraged by these events, several groups of private, semi-private, and public enterprises became active; a transition occurred from scientific interest in the metal-bearing lumps to commercial interest. But these pioneer developers faced a formidable task. Information about the minerals and their environment of deposition was inadequate; technology for mining them continuously was non-existent and very little was known about the adaptability of processing technologies for land-based ores to these minerals.

The mineral itself is a unique type. The lumps known as manganese nodules[1] are porous, concretionary objects of various sizes and shapes, found in thin discontinuous surficial layers on the floor of the ocean[2] and contain, in some cases, economically attractive quantities of nickel, copper, cobalt and manganese (and possibly molybdenum, vanadium and titanium). They occur at depths of 5000 m, hundreds of miles away from shore.

The environment in which the mining operations will take place is also unique. The terrain of the ocean floor is uneven, abounding in seamounts, hills, ridges, troughs, scarps, outcrops, boulders and other irregularities and obstacles. Generally, the sediments on which the nodules rest are soft, fine-grained, water-saturated, clay or ooze. The sea surface is affected by waves, ocean swells, currents, and sometimes storms and cyclones. Different types of currents are encountered at various depths of the water column. Such conditions present a very difficult environment for mining operations dependent on remotely controlled mechanical apparatus.

Over the last twenty years nearly half a billion dollars were spent in exploring the seabed for these mineral deposits, and in research and development of technology for mining and processing them. These initial undertakings were carried out by four multinational consortia composed of companies from the United States, Canada, the United Kingdom, Federal Republic of Germany, Belgium, the Netherlands, Italy, Japan, and two groups of private companies and public agencies from France and Japan. Three publicly sponsored entities from the USSR, India and China are also known to have undertaken some work.[3] A number of other countries also have interest in participating in future commercial operations.

The work of the pioneers over the last two decades has brought the process to a stage where, at the present time, it appears that ocean floor areas with potentially economic ore bodies have been identified and the basic technical feasibility of mining and processing technology has been established.

This volume presents a general state-of-the-art description of what has been accomplished so far and discusses what is to be expected next. Its aim is to provide a basic understanding of the process by which the technology for delineating nodule deposits and exploiting them has been developed, highlighting critical factors without undue technical detail. The analysis is based on information in the public domain. Much of the information on exploration, mining and metallurgical test results is proprietary; however, despite this constraint, the volume attempts to render a valid portrayal of (1) the geology of the deep ocean floor and the methods and procedures that have been used in prospecting and exploration, and (2) the physical problems of recovering nodules from the ocean floor and transferring them to transport ships, and the alternative equipment and methods that may ultimately be used in commercial mining.

Chapter 2 describes the mineral characteristics, and the depositional environment, factors that have had a considerable impact on the types of technology. Chapter 3 describes programmes that need to be carried out in order to reach the stage of making decisions about commercial exploitation of manganese nodules. Chapter 4 describes the methods and equipment used for manganese nodule exploration. Chapter 5 describes the three alternative mining systems being developed and tested. Chapter 6 is a broad assessment of the mining technologies. Following a discussion of alternative component and sub-system designs, a preliminary comparison is made of the different mining systems to shed light on their relative advantages and disadvantages. The technologies for transferring ore from the mining vessel to the transport vessel are dealt with in Chapter 7.[4] Chapter 8 sets forth some conclusions as to the technology presently being used for nodule exploration and mine tests and attempts to foresee the needs and probable development of more improved technology before commercial exploitation.[5]

Included in this volume are three appendices — the first gives a listing of patents for nodule exploration and mining technology; the second identifies possible suppliers of technology; and the third presents a bibliography. The lists of patents and technology suppliers as well as the bibliography, while not exhaustive, are comprehensive enough to offer a broad overview.

Chapter 2

Manganese Nodule Deposit

The process of converting manganese nodule occurrences to commercially exploitable deposits is shaped by the characteristics of the mineral and the environment of deposition.

NODULE CHARACTERISTICS

Manganese nodules occur on the floor of the ocean in a single layer. Rather than digging up the soil to recover the ore, as is generally the case in land-based mining, nodules have to be swept or scooped up. For gathering a large number of nodules, a wide area of the ocean floor needs to be covered. Nodules are of various shapes and sizes. An average nodule is slightly ellipsoidal with a diameter of 2.5–5 cm. Nodules vary in size from a few millimetres to many centimetres in diameter. Nodules are porous, containing water to the extent of one third to half their weight. Another characteristic associated with their porosity is their extreme fragility; they break easily and are crushed easily.

Manganese nodules contain various metals, among which nickel, copper, cobalt, manganese (and molybdenum, vanadium and titanium) are considered to be of economic interest. The grade of nodules (the content of the various metals of interest in a nodule expressed as a percentage of its dry weight) and the abundance of nodules (the weight of wet nodules per unit area of ocean floor usually expressed as kg/m^2) determine the amount of metals contained in the nodules *in situ* in a given area. Grades for potential economic deposits have been given in the general range of 1.1–1.6% nickel, 0.9–1.2% copper, 0.2–0.3% cobalt and 25.0–30.0% manganese. The range for abundance is indicated as 5–15 kg/m^2.[6] The grade of nodules in different areas of ocean floor and also within a particular area varies to a considerable degree. There are also variations in the abundance of nodules between and within areas of ocean floor.

ENVIRONMENT OF DEPOSITION[7]

The environment of deposition is characterized by ocean floor morphology, water depth and water column conditions, water surface conditions and distance from shore. The environment determines the operating conditions under which mining will take place.

Ocean Floor Morphology

Contrary to popular belief, the ocean floor where the nodules occur is not a smooth featureless plain: mountains, ridges, hills, scarps, troughs, basalt outcrops and boulders abound. The large seamounts range in height from 800 to 1500 m. The abyssal hills may be as high as 30 to 300 m, as long as 6 to 15 km and as wide as 2 to 5 km; the spacing of the hills may be as narrow as 30–60 m. Troughs can be 30–50 m deep, 250 m wide and 2 km long. The slopes of some of these troughs may be in excess of 30°.

Microtopographic features are variations in the ocean floor occurring over short distances, up to 100 m. Elevation changes of up to 1 m may be associated with these features. Other features may also be encountered including sediments that may be covered by pillow basalt, rocks and boulders of several metres in diameter, and narrow troughs, a few metres in depth.

The sediments on which nodules lie are fine-grained and of Bingham plastic type. The bearing or shear strength of the sediments can vary from area to area.

Water Depth and Water Column Conditions

Nodules have been known to occur in relatively shallow water depths; however, nodule deposits of commercial interest with requisite grade and abundance are found only on the deep ocean floor at depths between 3000 and 6000 m. Over a given area, local water depth can vary to a certain extent.

Water Surface Conditions

Wind, waves, swells and currents render the sea surface far from tranquil. The ship constantly undergoes various types of oscillations. Storms on high seas are not uncommon and extreme storm conditions and cyclones may occur occasionally.

Distance from Shore

Distances between the areas where mining may take place and the shore may be on the order of 500 to 3000 miles. This means that the market for the products and the support base for the mining system will be hundreds of miles away.

Chapter 3

Nodule Development Programme

Deep seabed mining is a new and emerging endeavour that is seeking to convert a unique type of mineral occurring in a unique environment into a source of metal to supply the world economy. From the perspective of mineral economics, this endeavour involves converting a mineral occurrence to a mineral resource and then to a mineral reserve. *A mineral occurrence* is one that is known to exist, but with no medium-term prospects as to its exploitability, technologically or economically. *A mineral resource* is an ore body that is not only known to exist, but that with improvements in technology and/or economic factors can be expected to be mined in the near future (up to twenty-five years). *A mineral reserve* is an ore body that can be worked under existing economic, social and political conditions.

The manganese nodule developers started with a mineral occurrence with no proven technology for exploiting it. Under these circumstances, the developers pursued an integrated programme which was geared to gradually reducing the uncertainties about geological and technological factors so that at the end, combined with an assessment of economic and socio-political factors, a decision could be taken about economic exploitation with acceptable risks. The sequence of activities is commonly termed the exploration sequence. The exploration sequence is outlined in Table 1 and the following sections elaborate on the sequence.

CONSIDERATIONS IN CHARACTERIZING ECONOMIC DEPOSITS

For a nodule mining venture to be economic, it is commonly felt that an amount varying from about 1.4 to 9.0 million metric tons of wet nodules needs to be mined annually for a period of 20 to 30 years. Within this range, the most often quoted figure is 3.0

Table 1. Exploration Sequence in Nodule Development

---

1.   REGIONAL APPRAISAL

Geological objective:	Select regional target areas.
Method:	Desk top appraisal and reconnaissance survey (high ship speed providing maximum reconnaissance data in minimum time expended).
Type of geological data:	Coarse grid bathymetry; indications of ocean floor hazards to mining activity, and indications of nodule grade and abundance.
Typical equipment:	Survey vessel (minimum 150′, long cruise endurance); articulated crane, geology lab, dark room, offshore navigation-satellite navigation; PDR, high resolution sub-bottom profiler, bottom sampling devices and cameras.
Skills:	Operations management; data evaluation and survey planning (marine geology, geophysical interpretation, cartography); ship operation and surveying; design engineering and related engineering.

2.   TARGET AREA REFINEMENT

Geological objective:	Provide data for refinement of target areas.
Method:	Closer grid bathymetric/geophysical survey; sampling and photographing in selected target areas (relatively high ship speed providing maximum data in minimum time expended).
Type of geological data:	Same as above, in selected target areas.
Typical equipment:	Same as above.
Skills:	Same as above.

3.   IDENTIFICATION OF PROBABLE ECONOMIC DEPOSIT(S)

Geological objective:	Provide sufficient data to select probable economic deposit.

Method: Closer grid bathymetric/geophysical survey and bottom sampling within refined target areas (slower ship speed accepted where necessary to obtain more quantitative, site specific data).

Type of geological data: Same as above plus more quantitative attempts at sampling refined target areas and obtaining information on seafloor characteristics.

Typical equipment: Same as above plus deeptow sidescan sonar and related support equipment (storage reels, wire, coaxial cable, etc.) and corers for sediment sampling.

Skills: Same as above, expanded capabilities to handle additional geophysical investigation and engineering tasks.

4. DELINEATION OF ECONOMIC DEPOSIT(S)

Geological objective: Provide sufficient geological data to allow delineation of an economic ore deposit (to be utilized in conjunction with technological, economic, and environmental data); prepare a mining plan suitable for the commencement of commercial mining.

Method: Closer grid bathymetric/geophysical survey and bottom sampling within probable ore deposit(s) where required. Emphasis on local variation of mineability. Survey interface with site specific mining and processing tasks.

Type of geological data: Same as above plus more quantitative sampling data and more detailed data on seafloor characteristics, ore mineralogic-chemical properties for processing decisions.

Typical equipment: Same as above plus accurate local navigation (satellite navigation buoy, and associated gear), and possibly dredges.

Skills: Same as above plus expanded capabilities to handle local navigation, expertise for geotechnical, mining and processing interface, and expertise for economic evaluation.

million metric tons of dry nodules, which is equivalent to about 4.5 million metric tons of wet nodules.[8] Assuming that an annual production rate of 4.5 million metric tons of wet nodules for 20 years needs to be maintained, the *mineable areas* in the mine site should contain *in situ* an amount of ore greater than 90 million metric tons of wet nodules, since not all the *in situ* nodules can be recovered.

The *mineable areas* will be defined in terms of three crucial factors — grade of nodules, abundance of nodules and seafloor characteristics. As noted in Chapter 2, grade, abundance and seafloor characteristics vary from area to area. Areas with low grade of nodules may not be considered worth mining currently. Likewise, areas with low abundance may be avoided. Areas with seafloor characteristics not amenable to first generation mining technology may also be eliminated.

Based on the average grade of nodules, a mine site can be divided into two sets of areas: (a) those areas where nodules have a grade higher than a cut-off level, and (b) areas where nodule grade is below cut-off (low grade areas). Based on the abundance of nodules in a given area, a mine site can be divided into (a) areas where nodule abundance is greater than a cut-off level, and (b) areas where nodule abundance is less than cut-off (low abundance areas). Finally, based on seafloor characteristics in a given area, a mine site can be divided into (a) areas where seafloor characteristics (slope, number and size of obstacles, sediment shear strength) are within an acceptable range, and (b) areas where seafloor characteristics are unacceptable. In this context, *mineable areas* will be defined as having a combination of grade and abundance above respective cut-off levels and acceptable seafloor characteristics. A mine site has to contain a sufficient number of mineable areas capable of supporting an economic mining venture.

As for mining technology, it has to be operable for about 300 days in a year for a period of 20 years, recovering about 15,000 metric tons of wet nodules daily. The size of the mining system and its sub-systems and components and/or the number of production units will be determined by this requirement. Mining technology will, of course, have to have the capability to tackle the physical operating conditions to be encountered in the mining environment. For example, the collecting mechanism will have capability for mobility within the constraint of the bearing strength of the sediments, capability for obstacle avoidance or survival, capability for sediment-water separation; the lifting mechanism will have the power to hoist the nodules, will have the capability of withstanding static and dynamic forces encountered in the water column, etc.[9]

EXPLORATION SEQUENCE

The approach to ocean exploration for economic deposits has borrowed heavily from methods used traditionally for land exploration. Bailly[10] shows a generalized sequence of exploration activities for land-based mineral deposits. It is, generally, the same approach as that shown by the deep sea mining companies. Tasks are carried out in sequential stages — analysis of a specific task performed at one stage leading to a decision as to whether to undertake the next task and when to do so.

Economic considerations play an important role in the planning and implementation of the overall exploration sequence. Efforts must be made to maximize the time and capital expended in identifying and exploring the most promising target areas of occurrence and in selecting and refining the most promising design concepts of mining technology.

In the initial stage of the sequence and on the geological side, existing data and information about the mineral and the environment of deposition are appraised for the purpose of establishing commercial viability. Information about some of the physical factors exists in the public domain to a certain extent. Information available in the public domain was obtained mainly from oceanographic cruises and other information activities carried out by academic institutions and government agencies. To a lesser extent, some of the information was obtained from the exploration activities of private industry. Most of these data, however, were gathered for purposes other than establishing the commercial viability of an area and they need to be augmented by site specific data to achieve the latter objective.

Information about sea surface and water column characteristics is generally available. The most important sea surface parameters are waves — both wind waves and swells. In general, existing wave data are reliable for assessing the effects of nominal weather conditions. Precise data about the frequency and duration of extreme weather conditions are not generally available. However, the existing data allow general conclusions regarding the geographical and temporal variability of storm and hurricane activity.

The availability and usefulness of the public data about macrotopography vary from region to region, but for particular regions, data about large topographical features — fracture zones, seamounts, ridges, troughs etc. — are considered sufficient. Public data about microtopography and obstacles are meagre, although the newer sonar systems recently in use by oceanographic institutes have the potential of augmenting public data about these factors.

The adequacy and reliability of the publicly available data on the nodule deposits for the purpose of general resource assessment have been discussed in Volume 1.* However, existing information for the purpose of delineating potentially economic deposits is of limited value. Data in public domain can point to the regions which are worth further investigation. For example, the public data available in the early 1970's indicated that the Clarion-Clipperton Zone in north-east Pacific Ocean was the most promising region and that indication has been reinforced by the findings of the pioneer developers in that zone.

After an appraisal of existing data and indentification of large regions worth investigation, the exploration sequence proceeds with reconnaissance survey of the regions. The purpose is to select smaller target areas within the large regions worthy of more detailed investigations.

In the next step, the target areas are surveyed in further detail to provide a larger data base, which allows further exploration decisions at a higher level of confidence. The end results are target area refinements — some target areas are rejected, some are retained for future consideration and some promising ones are selected for further investigation. Regional appraisal and target area refinement usually constitute the prospecting stage.

Additional surveying is then performed in the selected target areas to provide a data density sufficient to allow the identification of probable economic deposits. The evaluation of these data further refines the remaining target areas, allowing future exploration efforts to focus on deposits that have the most promising economic potential. In the next stage, these deposits are further investigated with the objective of delineating an economic deposit. In the final stage, the activities provide sufficient data to allow evaluation of the economic deposit, delineation of a mine site and preparation of a tentative mining plan for the initial period of commercial operation.

In order to arrive at the desired exploration objectives, large quantities of data must be collected systematically from vast areas of ocean floor, and evaluated in an efficient manner. Because it is impossible to sample every square metre of a selected target area

---

Assessment of Manganese Nodule Resources

and because of the variations in grade and abundance, exploration activities rely on statistical and, in particular, geostatistical methods of treating the data and make decisions based on probabilities.

In target area exploration, the usual approach to sampling design, following practices in land-based mining, involves the gridding of areas to be surveyed into square cells. Usually, a hundred or more cells with 3–10 discrete samples in each cell is considered a reasonable number. However, the number of cells and the number of samples taken per cell must ultimately reflect judgements made concerning the variation in grade and abundance and the accuracy expected to be achieved by the statistical methods. Statistical analysis of data collected within each cell is carried out on the assumption that the samples are representative of the given cell area at an assigned level of confidence. Further statistical treatment of all cells in an explored area is then used in immediate and future exploration activities.

Care must be taken during grid planning to ensure a cross-pattern of ship tracks. Ship tracks in only one orientation can introduce serious bias into the evaluation of geophysical data by enhancing physical features oriented normal to ship track direction.

Target area exploration proceeds from coarse grid survey to increasingly finer grid surveys so that sufficient bottom sample data and data about seafloor morphology are generated for the delineation and evaluation of an economic deposit and, ultimately, delineation of a mine site and preparation of a mining plan.

Figure 1 illustrates the progression from a 30 nautical mile grid spacing down to a 0.5 nautical mile grid spacing. In this example, initial statistical evaluation of reconnaissance data may have indicated 2 areas (A and B) worthy of further, more detailed exploration. Upon closer inspection (10 nautical mile grid) Area B may have been rejected on the basis of unsatisfactory nodule or seafloor characteristics. Area A is further explored at a higher data density and if statistical evaluation of the area appears promising, additional exploration may be carried out at a 1 or 0.5 nautical mile grid spacing.

Similarly, on the technological side, research and development of manganese nodule mining technology starts with an examination of the pool of existing technologies and an assessment as to the extent to which they can be adapted. Of the range of

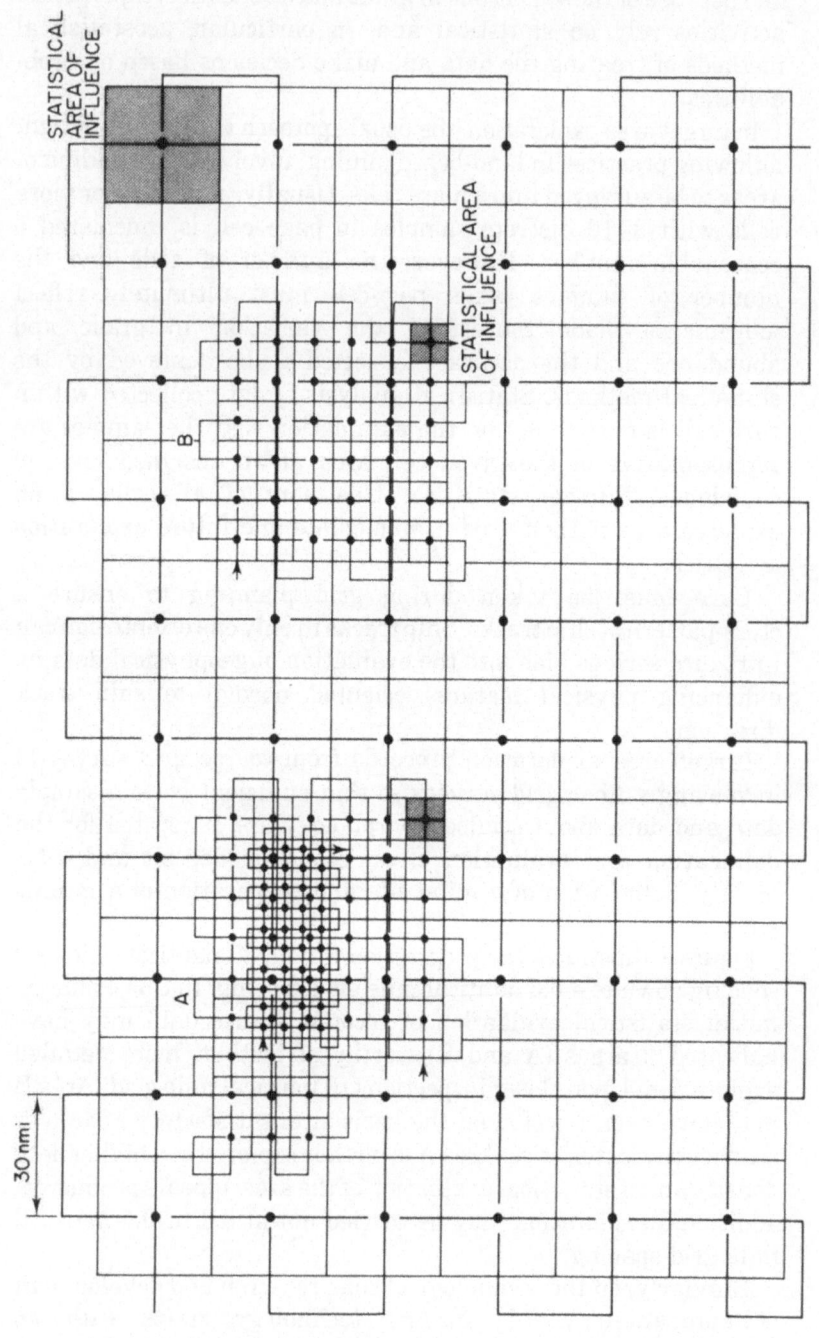

Fig. 1. Progressively closer gridding of exploration areas.

available technologies, the most functionally similar is that used for offshore dredging. Currently, the deepest waters in which dredging technology is applied are about 100 m. Nodule mining will be done at depths which are more than an order of magnitude greater than this. Offshore drilling technology is another technology which provides lessons and scope for adaptability. Fixed platforms operate in depths of up to 300 m, and guyed towers and tension leg platforms are being developed to operate in depths of 300–700 m and 700–1000 m, respectively — still a long way off from the deep sea. However, these technologies have relevance to deep sea mining in respect of the surface platform, pipe handling, materials handling, structural design, approaches to dealing with surface conditions, station keeping and logistics. Research drill ships have operated in depths and with pipe lengths comparable to deep sea mining. The nodule mining ship, however, has to operate for a much longer period; the mining system has to be much more mobile, requiring capabilities for withstanding dynamic forces on the pipe and the collector; the collector has to tackle much more varied bottom conditions; and the coordination between the mining ship and the transport ship needs to be maintained. Still the technology of the research drill ship is relevant to deep sea mining technology in the areas of navigation, position keeping, ship-pipe interface, pipe handling, and pipe design.

Nodule mining technology cannot be developed by incrementally modifying existing technologies, nor is it a matter of putting together off-the-shelf sub-systems and components from the pool of existing technologies. New concepts, sub-systems and components have to be developed. A systems engineering approach is called for since modification in one sub-system or component has implications for other sub-systems and components and thus for the whole mining system.

After the assessment of the existing pool of technologies, gaps are identified which need to be filled by innovative technologies. The development of an entirely new system composed of interrelated sub-systems and components, some of which are new, must first have an over-all conceptual design. A wide array of design concepts for the whole system and for the new sub-systems or components is examined through preliminary engineering studies to select design concepts which look promising.

Selected design concepts are then refined; a model or pilot scale component, sub-system or system is fabricated or integrated and tested. Depending on the evaluation of the design concepts, tests of components may be done in a simulated ocean environment in a laboratory, or in shallow water or the deep ocean. In order to establish its basic technical feasibility, the whole system may have to be tested in the deep ocean.

Based on the pilot test results, further phases of design refinement and testing are carried out. A large scale prototype testing in the deep ocean for a relatively long period may be necessary in order to ultimately settle on the design and fabrication of the commercial scale system.

A crucial characteristic of the exploration sequence is that the geological and the technological activities are carried out in an interactive manner. Each stage of the technology R & D process identifies further data needs and the results of each geological activity are used in subsequent stages of the technology R & D process and vice versa. For example, the results of reconnaissance throw light on the characteristics of the ore and the environment of deposition; design engineers use this information in refining their design concepts which in turn, influence the type of information to be collected through target area exploration. Using the information collected through target area exploration activities, engineering determinations are made of the techniques and equipment that can work on a potential economic deposit and the processing routes that are most suitable for the ore. The general nature of the interactive process is illustrated in Fig. 2.

It is maintained by some researchers that nodule mining technology is essentially deposit or site specific and technologies for different deposits or sites vary not in degree but in kind. However, it is likely that in the initial stage there is a trade-off between:

(a) designing a technology applicable more universally, which will involve less information collection costs and will offer flexibility regarding choice of site but which at the same time will reduce performance locally at a specific site and require further work to develop a mining plan at that site, and

(b) designing site-specific technology which will increase performance at the given site and will offer an associated mining plan but which, at the same time, will involve more information costs and will allow less flexibility.

It should be noted that the exploration sequence is not necessarily a linear one; it is more likely than not that the process

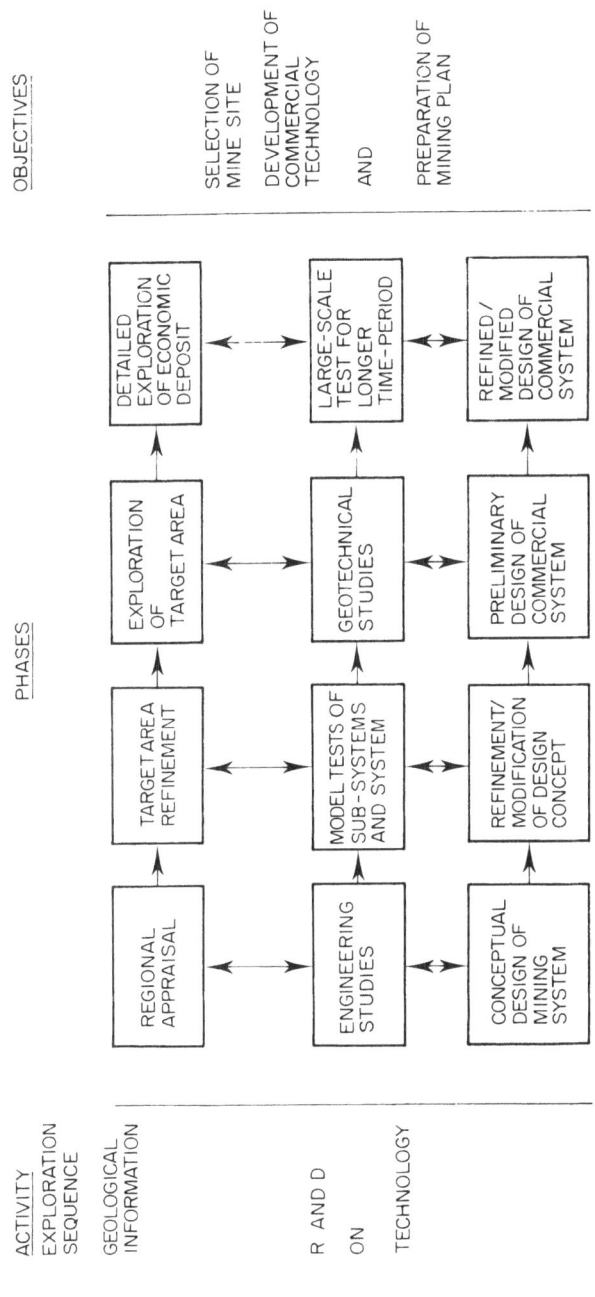

Fig. 2. Interactive process of technology development and information collection in exploration sequence.

is iterative. On the geological side, a target area which seemed promising initially may have to be discarded after further investigation and exploration activities transferred to a target area which was originally retained. Similarly, development of technology operable in a physical environment about which not much was previously known has to proceed in an iterative fashion. Initial design concepts are developed based on existing knowledge. As more information about the physical environment becomes available, the design concepts are modified. Some design concepts which seemed attractive initially may prove to be blind alleys after an analysis of the test results and with a better understanding of the nodule deposits and seafloor characteristics; at times, entirely new design concepts may appear more viable.

PRESENT SITUATION

At the present time, judging from the information available in the public domain, it can be concluded that the developers have completed the prospecting stage and seem to be ready for the next stage of the exploration sequence. They have appraised large areas of ocean floor — Clarion-Clipperton Zone in the north-eastern equatorial Pacific Ocean and a region in the Indian Ocean. They have identified target areas within these large regions, which have some probability of containing ore capable of supporting an economic mining venture. Engineering studies, design conceptualization of system, sub-system and component and pilot testing have resulted in identification of design and equipment with some probability of being applicable on a commercial scale. Most of the routes required to process manganese nodules into metals of interest have been developed.

Thus, a progression seems to have been made in the exploration sequence from a mineral occurrence to a mineral resource. The next stage will involve a progression from a mineral resource to a mineral reserve.

Chapter 4

Technology for Manganese Nodule Prospecting and Exploration

METHODS AND EQUIPMENT

Prospecting and exploration devices can be of three types, broadly: bottom sampling devices, visual devices and acoustic devices.[11] Figure 3 makes a schematic presentation of various devices. Bottom sampling devices, namely grab samplers (both the conventional type attached to the ship and the free-fall samplers), piston corers and box corers, and dredges have been described in Volume 1, as have the visual devices, namely still cameras and TV cameras.

Acoustic devices include echo-sounders, sub-bottom profilers, sidescan sonars and other kinds of sound reflecting and refracting devices. The basic idea of acoustic devices is to generate sounds to be reflected by various kinds of materials and surfaces, monitor them and analyse them to obtain relevant information about the materials and the surfaces.

The main purpose of prospecting and target area exploration is collection and analysis of information about manganese nodule deposits and the environment of deposition. Certain factors, e.g. the characteristics of the nodule deposit, in particular, the abundance and grade of the nodules and seafloor macro- and micro-topography are crucial.

Information is used as input to the two interrelated objectives of the exploration sequence, delineation of an economic deposit and development of technology. The relevance of the information about the sea-bottom, water column and sea surface to each of these two goals is shown in Table 2. It should be pointed out that the degree of relevance can vary from critical to marginal, and a preliminary effort has been made in the table to identify the critical factors (denoted by an asterisk).

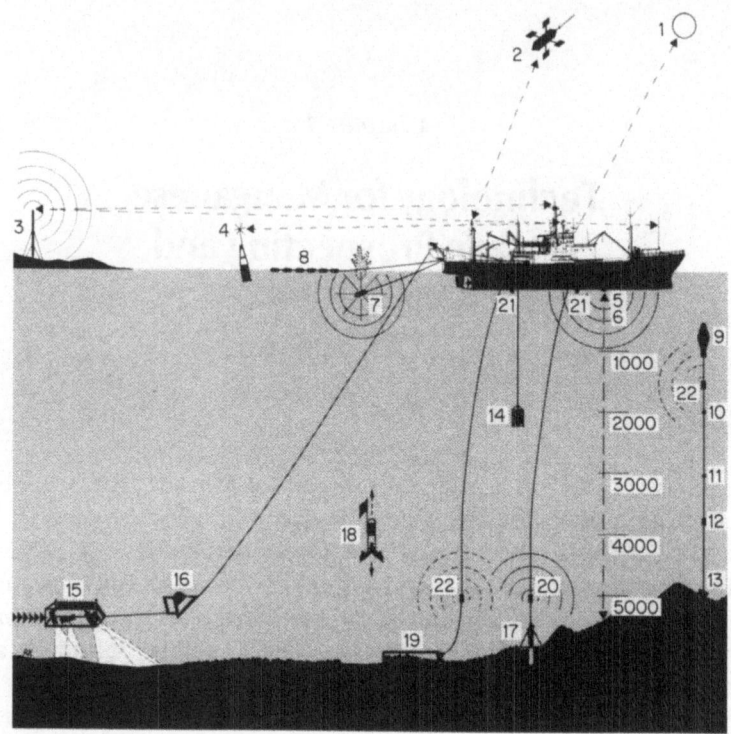

Fig. 3. Various deep sea prospecting and exploration devices. (Courtesy of Arbeitsgemeinschaft Meerestechnischgewinnbare Rohstoffe, Federal Republic of Germany.)

KEY:

Navigation
1 stars
2 satellites
3 radio navigation
4 navigation buoy (transponder/radar)

Bathymetry
5 narrow beam echo sounder, sediment echograph
6 various depth recorders including precision depth recorders (PDR)

Sub-bottom
7 sub-bottom profilers (SBP)
8 streamers

Water column
9 underwater measuring chain with localizable buoy

10 current meter
11 thermometer
12 water pressure gauge
13 cut-off anchor
14 bathysonde (continuous measurement of temperature, salinity, sound velocity, pressure)

Nodule characteristics
15 deep tow platform
16 stabilizing platform
17 corer for sampling sediment with nodules
18 freefall sampler
19 drag dredge

Localization of launched survey gauges
20 pinger
21 hydrophone
22 transponder

Table 2. Information about Physical Factors as Inputs to Delineation of Economic Deposits and Development of Mining Technology

Physical Factors	Development of Mining Technology	Delineation of Economic Deposits
1. Surface		
1.1 Atmosphere		
1.1.1 Winds*	Wind forces including gusts affect ship motions, and are especially important in design of pipe handling system.	Important in regional appraisal, but relatively unimportant in identifying probable deposits within given target areas, since there is little variation in these factors over a given area.
1.1.2 Humidity	Significant in corrosion, operation of electronic gear, and crew comfort.	
1.1.3 Visibility	Significant in navigational and radar requirements.	
1.1.4 Temperature	Significant in operation of electronic gear and crew comfort.	
1.2 Water Surface		
1.2.1 Waves and Swells*	Extremely important in design of collector and pipe handling equipment, mining system dynamic environment, definition of operating plans, estimates of system availability and survivability.	
1.2.2 Tides	Not important in design in that effects are small in relation to other vertical displacement. Useful in estimates of navigational error.	
1.2.3 Currents	Important in estimates of pipe loads, ship propulsion requirements, and operating plans.	
1.2.4 Fluid Properties, including specific gravity, temperature, salinity, and oxygen content	Significant in estimates of system drag, corrosion environment, and pump power requirements.	

Table 2 (continued)

Physical Factors	Development of Mining Technology	Delineation of Economic Deposits
2. Water Column		
2.1 Currents	Significant in pipe drag; variations contribute to pipe vibrations.	
2.2 Salinity	Variations contribute to internal wave effects; however, these will be negligible. More important in corrosion.	
2.3 Temperature	Useful in any estimate of slurry temperature at the surface.	
2.4 Specific Gravity	Important in estimating pump power requirements and buoyant forces on the system. However, variations from nominal values cause secondary effects on system loads.	Same as above.
2.5 Oxygen Content	Useful in estimates of corrosion rates.	
3. Bottom		
3.1 Bathymetry or Macrotopography*	Extremely important in establishing system pipe length, and depth accommodation requirements and mining plans which comply with depth accommodation capability. Contributes to collector control requirements, especially side-slip and operating speed variations.	Significant in that depth variation determines sub-area exploitable with a given pipe length.
3.2 Microtopography*	Extremely important in establishing collector dynamics which influence collection rate and collection rate variations. Strongly influences running gear and collector control requirements.	Has significant impact in defining mineable areas.
3.3 Obstacles*	Extremely significant in influence on collector survivability requirements, collector control requirements, navigation requirements, and survey requirements. Significant factor in mineable area estimates and an essential consideration in mining plans.	Significant factor in defining mineable areas.

3.4	Sediment Physical Properties*	Nominal values are important in setting system requirements. The intrinsic tendency of these properties to be highly variable means system design and operational approaches must generally be insensitive to these variations.	Variations from site to site have some impact through collector design and operating requirement.
3.5	Nodule Supply		
3.5.1	Abundance and spatial distribution of abundance*	Extremely important variable in production estimates, and collector size and speed requirements. Variability in concentration is important in establishing peak power and nominal slurry concentration requirements.	Abundance is dominant consideration in defining economic deposit.
3.5.2	Grade*	An important factor in mining plans and shipboard processing requirements.	Extremely important in defining economic deposit.
3.5.3	Size Distribution*	Important in the sizing of dislodgers, crushers, if any, and lift pipe. Significant in the design of shipboard separation equipment.	Significant in that size distribution influences the efficiency of dislodgers, pipe capabilities and shipboard separation capabilities.
3.5.4	Embedment Depth	Significant in assessing nodule dislodgement and pick-up requirements.	Very important in estimating nodule concentration from population.
3.5.5	Fragility*	Important consideration in attrition rates of nodules. Significantly impacts on collector separation requirements, pump power requirements, and separation requirements on the surface.	A consideration, but likely to be of second order in site to site evaluation. May provide an index to beneficiation processes inherent to proposed mining methods (i.e. removal of clay and detritus from nodule interiors).

Table 2 (continued)

Physical Factors	Development of Mining Technology	Delineation of Economic Deposits
3.5.6 Density	Important in establishing pump power requirements and system flow rates.	A consideration, but likely to be of second order in site-to-site evaluation.
3.6 Water Turbidity	Nominal values are important in the design and operation of any optical scanning equipment.	Site-to-site variations are not likely to be a significant factor in site selection.
3.7 Sub-bottom	Not of direct concern in system design and operation.	Has considerable value in the reconnaissance phases of exploration as implied by the correlation of acoustic properties of surface sediments with nodule abundance. Also important in prediction of hazards to mining through indications of proximity of acoustic basement (rock) to the seafloor.

(Reprinted with permission from Science Applications, Inc. and United States Bureau of Mines)

Bottom Sampling Devices

For obtaining actual samples of manganese nodules and sediments from the bottom of the ocean, three types of bottom sampling devices are used — grab samplers, corers and drag dredges. Various designs of each of these devices are available. As for corers, three types are distinguishable — gravity corers, piston corers and box corers.

The sampling devices can be lowered to the bottom from the ship with a wire held from the ship and retrieved by pulling the wire up. They can also be dropped from the ship, unattached to any wire; these free-fall devices are equipped with disposable ballast materials and buoyancy chambers. The buoyancy mechanism, by discarding ballast, causes the devices to return to the sea surface where they are retrieved by the ship. Because of the relative lightness and small size of these devices, the free-falling method has been used so far for grab samplers, gravity corers and piston corers only.

Grab samplers have two jaws or scoops. They drop to the ocean bottom with the jaws open and, upon hitting the bottom, the jaws close retaining nodule samples. The jaws are usually made of metal, the container is usually a heavy mesh or canvas bag. Grab samplers can be built to various size specifications but normally they do not cover more than a square metre and weigh less than 75 kg.

While grab samplers scoop up the nodules by closing their jaws, corers are long hollow cylinders which penetrate the ocean bottom and encase a core of the bottom sediment. Gravity corers penetrate the bottom with the force of their own weight. Piston corers have a piston inside and penetrate the bottom by using hydrostatic pressure from the piston. Gravity and piston corers cover only a few square centimetres of the ocean floor. Box corers are bottomless metal boxes which push their four thin walls into the bottom sediment by their weight (generally between 350 and 750 kg). A sharp blade or spade rotates through the soil cutting and encasing a chunk of bottom soil along with the nodules at the top. Box corers can be of various sizes, but they cover about a square metre of the ocean floor with a depth of 25 cm.

Drag dredges are open-mouthed containers (boxes, pipes or bags) that are dragged along the ocean floor for the purpose of collecting a relatively large number of nodules. The open mouth is about 2 m wide and 35 cm high; the body of the container itself may be about 1 m or longer.

Visual Devices

Visual surveys of manganese nodules on the ocean floor and the surrounding topography are carried out by still, movie and TV cameras. When used independently, cameras can be mounted on cable-supported frames or lowered to the sea floor on cable-suspended devices towed slightly above the ocean floor; still cameras can be placed on free-fall devices.

Photographs by conventional still cameras are taken either by an automatic triggering mechanism snapping at preset intervals or by a suspended triggering device activated by contact with the bottom. Strobe lights provide illumination for the camera. Because of the absence of light at extreme depths of the ocean, and the difficulty of transmitting a strong light source there, lighting is limited. Also any light tends to scatter in water. For these reasons, photographs must be taken from a close range and thus cover only a small area of the ocean floor.

Movie cameras can be used in almost any situation where still cameras are used, but the requirements of a continuous light source, close operator control and large amount of film have restricted the use of movie cameras.

The underwater television camera has several advantages over still and movie cameras. The greatest advantage is in obtaining pictures instantaneously. Also, it needs no film and the pictures are recorded automatically on magnetic tape. However, the TV camera has limitations, too. It is expensive and requires a special cable and winch, as well as costly monitoring equipment. The camera covers only 15 m^2, and takes a long time to survey a large mine site. Also if an ocean floor obstacle is not shown on the screen, the shipboard monitor may not be able to stop the camera from colliding with it. Because of loss in the power sent over several miles of cable, a high light level cannot be achieved and, as a result, low light level cameras with poor resolution are used (fiber-optic cables may help to remedy this problem in the future). Because of the relative lack of light these cameras must stay very near to the ocean floor. To avoid the blurring that can result when an object is filmed from close up, the camera has to be towed very slowly, taking up expensive ship time. The camera also tends to hit the ocean floor often because of the rising and falling of the ship.

Acoustic Devices

Unlike high frequency light waves which are dispersed over distances, sound waves, which have a low frequency, can be transmitted easily in the ocean and acoustic devices are widely used for exploration purposes. There are several types of acoustic devices. Of these, the echo sounder (down-looking sonar) performs the basic task of determining water depth, the sub-bottom profiler studies the sediment layers and sub-seafloor features, and the up- and side-looking sonars serve to increase the area of the ocean floor that can be charted. Experiments with acoustic imaging devices, which are designed to visually reproduce small features on the ocean bottom, have been successful to some extent.

There are various types of echo sounders. The echo sounder is a down-looking sonar, mounted on the keel of the ship, which emits a sound pulse. It takes a certain amount of time for the sound pulse to reach the ocean floor and its reflected echo to return to a receiver on the ship. The length of time determines the distance to the ocean bottom. The speed of sound in water, which changes with water temperature and salinity, needs to be taken into account in order to make this calculation.

Given the water depth, delineation of small features and the determination of minor slopes requires a narrow sound cone which increases resolution of bottom features. Echo sounders usually have a wide sound cone; for example, a precision depth recorder (PDR) which is one type of echo sounder has a sound cone width of about 30°; the cone is usually rendered wider because of the rolling of the ship. Echo sounders, thus, describe seabed macrotopography well, but they cannot delineate smaller features of topography and minor slopes.

New echo sounding devices have been developed in an effort to achieve narrow sound cones. Gyro-stabilized sonar transducers eliminate the effects of ship rolling and have a much narrower sound cone than PDRs. These, however, cannot discriminate between objects smaller than 175 m in 5000 m water depth.

In achieving narrow sound cones, however, an additional consideration is the fact that with a narrow sound cone, the area of the ocean floor covered is small and thus, long ship time and consequent costs are required to obtain information about a given size of the total area to be explored. The use of an array of narrow-beam echo sounders has been developed to achieve the dual objectives of narrower sound cones and wider coverage of ocean floor area. The civilian version of this multi-beam system, called the SEABEAM system, has, for example, an array of 20 echo-sounders, each with a sound cone of $2\frac{2}{3}°$. The reception mode can be contained within a smaller cone than the addition of the cones of the individual echo-sounders, thus effectively achieving the results of narrower beam; at the same time, the total cone width is broad enough to cover a relatively large area of ocean floor. This system is capable of resolving features on the ocean floor that may be as small as 20 m wide and 5 m high in ocean water depth of 3500 metres.

Sub-bottom profilers (SBP) are similar to echo-sounders, but they use a lower-frequency, higher-energy sound pulse for bouncing acoustic echoes off sediment layers beneath the ocean floor. These are useful in investigating the structure of the sedimentary layer making up the ocean floor and the surface characteristics of the ocean floor itself on which nodules may lie.

Side-scan sonars can look sideways, as it were, instead of looking down or up only. The sound cone has both a vertical width (about 20–40°) and a horizontal width (about 1–4°). The fan-shaped sound pulse along with the narrower horizontal cone are useful for the location of relatively small topographic features as well as for distinguishing major differences in sediment type. The products of the side-scan sonars are similar to aerial photographs; they are, however, much more difficult to interpret. Newer side-scan sonar systems, e.g. SEAMARK I and II produce better images and cover wider areas in less time.

Multiple Instrument Platforms

Each exploration device serves a different purpose and provides a different type of information (Table 3). Therefore, it is useful to operate a number of instruments in an exploration cruise. These instruments must be used in exactly the same area of the ocean floor in order to characterize it properly and fully for mining purposes. Sending down one instrument at a time wastes much ship time. For this reason, ocean miners are looking increasingly towards multiple instrument platforms to save time and money on nodule exploration. These platforms, commonly known as deep-tow fish, are lowered by a cable into water slightly above the ocean floor, where they can record several types of information simultaneously or sequentially by using the instruments one after another without having to resurface between measurements.

Table 3. Uses of Prospecting and Exploration Devices

Device	Measurements and Corollary Measurements (CM)* Required	Operating Characteristics	Performance Measures**	Performance
Free-Fall Corer	– Sediment properties, excluding direct measurement of engineering properties	– Free-fall and return with sediment sample in 1.5–2 hrs. – Maximum core length 1 m – Weight approx. 90 kg	– Edge effects and sample disturbance – Depth of penetration – Length of core recovered – Drift characteristics – Ship time required per sample	– Sample area 6 cm in diameter and up to 1 m long, dependent on physical properties of sediment
Free-Fall Grab Sampler	– Nodule grade – Nodule abundance – CM-position	– Free-fall and return with nodule sample in 2–4 hrs. – Weight approximately 75 kg	– Sampling speed – Drift characteristics – Precision with which sample area is known – Collection efficiency and variability – Size of sample – Positional accuracy obtainable – Ship time required per sample	– 0.2 m² sample area – Correction factor error – Positional accuracy about 450–1800 m – Usefulness greatly enhanced if equipped with a still camera
Box Corer	– Nodule abundance – Nodule grade – Nodule population – Nodule size distribution	– Lowered and recovered by winch while ship is stationary	– Accuracies – Edge effects – Sample disturbance – Operating speed	– Abundance accuracy better than 1% – Grade accuracy (limited by assay technique)

Table 3 (continued)

Device	Measurements and Corollary Measurements (CM)* Required	Operating Characteristics	Performance Measures**	Performance
	– Sediment properties – Nodule dislodgement forces – Nodule embedment depth – CM-position		– Positional accuracy obtainable – Sample area – Ship time required per sample	– Speed 2 hrs./sample – Position within about 150 m – 0.06–0.75 m² sample area
Drag Dredge	– Average grade over tow length – Large sample for other studies	– Towed from surface ship	– Capacity – Efficiency	– Two-metric ton capacity – Tow efficiency
Towed Fish Mounted Camera	– Nodule population – Obstacle detection – Obstacle characterization, especially with stereo photos – CM-position – Camera height above bottom	– 6–9 m frame width – Speed 0.5–2 knots – 9 m above bottom approximately – 5000 photos/camera run – Strobe lighting and batteries	– Resolution – Area covered/unit of ship time	– 35 mm (600 line pairs resolution) – 70 mm (3000 line pairs resolution)
Towed Fish Mounted TV	– Nodule population – Obstacle detection	– 2 m frame width – Speed 0.5–2 knots	– Resolution – Transmission rate	– 260 line pairs

	— Obstacle characterization, especially with stereo photos — CM-position — Camera height above; bottom	— Power cable required — Unlimited continuous imaging — Approximately 10 m visible at one time — Operated at nearly uniform height above bottom		— Area covered/unit of ship time
Free-Fall Still Photography	— Nodule population — CM-position	— Free-fall and return 2–4 hrs. — Camera tripped on bottom contact	— Photo position precision — Resolution — Area of photograph — Ship time/photo	
Precision Depth Recorder (PDR)	— Bathymetry — Macrotopography — CM-position — Crude measurement of nodule abundance possible	— Operated from surface ship — 30° sound cone — Ship speed not limited	— Profile m/sec. of ship time — Profile power transfer function or minimum detectable profile wavelength with specified accuracy	— Profile wavelengths greater than about 150 m detectable — Profile obtained at ship speed — Does not detect slopes < 10°
Sub-bottom Profiler (Near Surface Application)	— Sub-bottom profile — Indication of nodule abundance — Indication of obstacles to mining — CM-position	— Hull-mounted or towed 3–15 m below sea surface — Ship speed 0.5–12 knots	— Profile m/sec. of ship time — Profile power transfer function or minimum detectable profile wavelength with specified accuracy	— Depending on operating mode down to wavelengths less than 3 m — Profile produced at 0.3–1 m/sec.

Table 3 (continued)

Device	Measurements and Corollary Measurements (CM)* Required	Operating Characteristics	Performance Measures**	Performance
Sub-bottom Profiler (Deep Tow Application)	– Sub-bottom profile – CM-position – Indication of nodule abundance possible – Indication of obstacles to mining	– Towed 20–100 m above bottom – Ship speed 0.5–2 knots	– Profile m/sec. of ship time – Profile power transfer functions or minimum detectable profile wavelength with specified accuracy	– Depending on operating mode down to wavelengths less than 3 m – Profile produced at 0.3–1 m/sec.
Side Scan Sonar	– Obstacle detection and characterization – Microtopography – Possible nodule abundance measurement – CM-position	– Sound cone vertical with 20–40° – Sound cone horizontal with ¾–1.5° – Towed 20–100 m above bottom – Speed 0.5–2 knots	– Probability of obstacle detection – Obstacle height and slope estimation accuracy – Area rate (area/unit of ship time)	– Performance depends on many design and operating factors

* "Corollary measurement" is one which must be made at the same time as the principal measurement made with the device.

** Basis on which instrument performance can be assessed.

(Reprinted with permission from Science Applications, Inc. and United States Bureau of Mines)

There are several types of multiple instrument platforms that have been used. For example, the Scripps Institution of Oceanography operates a multiple instrument platform which is equipped with up-, down- and side-looking sonars, a sub-bottom profiler, a high precision magnetometer, a precision temperature measuring device, cameras, strobe lights and a television system. A supporting coaxial cable relays information to and from the fish. The main advantage of the system is its ability to provide measurement of nodule abundance within a continuous 1 km track of ocean floor.

The Unmanned Deep Ocean Survey System (UDOSS) was designed for the US Geological Survey by the Jet Propulsion Laboratory. The system is composed of three sub-systems:

(a) the fish itself which is mounted with sonars for the purposes of navigation, obstacle detection and geological studies, still cameras, underwater TV system, lighting for the photography, magnetometers, and pressure, temperature and conductivity measuring devices;

(b) the shipboard data processing sub-system which has the capability to display live television and acoustic surveys as well as previously recorded data on a shipboard console, it can also control and monitor the condition of all of the equipment on the platform; and

(c) 10,000 m of electrical cable that supports the platform, conducts power to it and transmits data from it directly into the shipboard console.

The Deep Sea Survey System (DSS-125) developed by Hydroproducts is a commercially available multiple instrument platform put on the market in 1974. The basic system comes with: low light level TV, thallium iodide lights for illumination, spotlights to determine height above ocean floor, electronic compass, 70 mm camera that can be controlled from the surface vessel, a strobe light; a depressor platform to keep the instrument platform at a low-angle slope in relation to the ocean floor by weighing down the cable in front of it; and shipboard control console to regulate the equipment on the fish and 7500 m of 21 mm armoured cable capable of relaying information. Optional equipment includes side-scan and forward-looking sonars for obstacle avoidance, sub-bottom profiler and acoustic devices for positioning.

The sampling devices and the optical devices can be used for spot-checking while acoustic as well as optical devices are suitable for deep towing. In both cases, as long as the devices are wire-held from the ship, the exploration process can be time-consuming, and thus expensive, since the ship speed is limited due to cable-towing considerations. One way to increase ship speed is to eliminate the cable and obtain information through remote sensing methods. Acoustic devices are amenable to remote sensing methods. Acoustic devices can be mounted in the ship or towed shallowly near the ship. Proper analysis and interpretation of data about the ocean floor obtained from these devices, correlated with the sound reflectivity of nodules, can result in fairly useful indications about nodule occurrence and abundance. Several such remote-sensing systems have been developed and are under development.

Figure 4 lists the components of prospecting and exploration technology.

ASSESSMENT OF TECHNOLOGY FOR PROSPECTING AND TARGET AREA EXPLORATION

It should be recalled that the tasks to be carried out and the objectives to be achieved are different in the prospecting and target area exploration stages of the exploration sequence. The same basic equipment can be employed in both the stages; however in the target area exploration stage, more detailed information, a larger number of bottom samples obtained according to a planned pattern and mapping of particular seafloor areas, as detailed as possible, are required.

In the prospecting stage of the exploration sequence, reconnaissance is carried out over a wide region with the objective of identifying smaller target areas for further investigation. Indicative rather than detailed data from a particular area are required about nodule characteristics and seafloor morphology. There is, thus, a need for over-all high ship speed which allows minimum time expended for maximum area covered at a lower level of confidence. Precision depth recorders can generate data which allows elimination of areas with macrotopographical features. Still cameras and short runs of TV cameras can give indications of coverage. The bottom sampling locations can be random and widely dispersed. The orientation of the ship track need not be planned carefully.

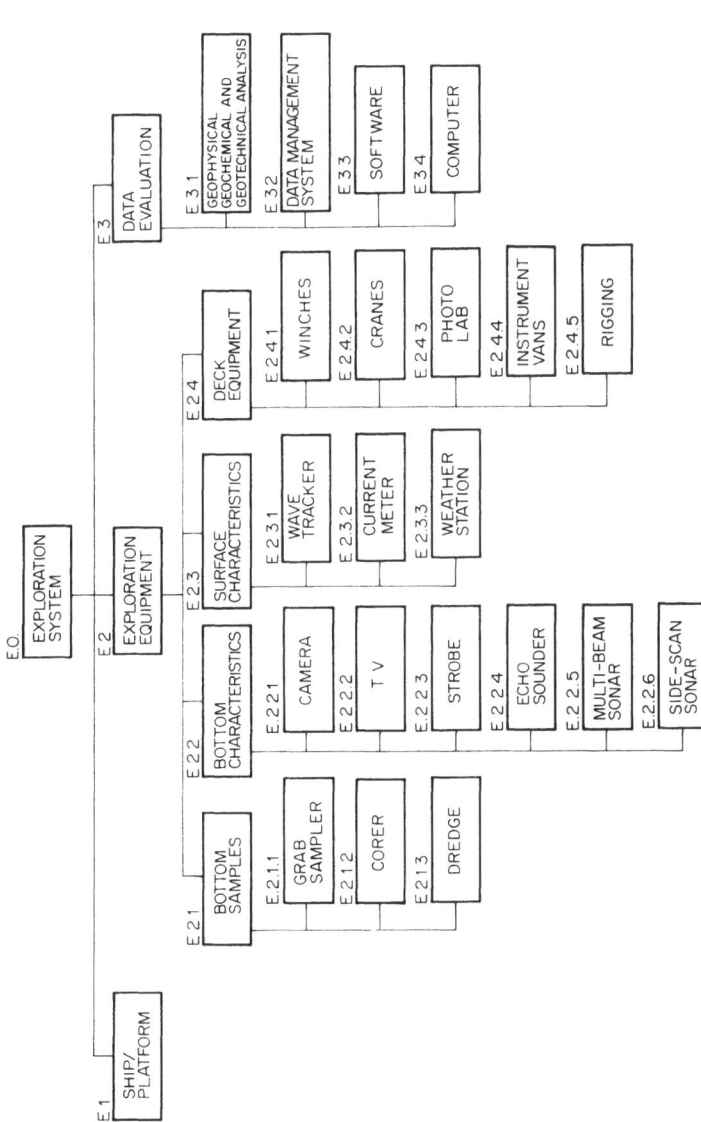

Fig. 4. Prospecting and exploration technology component breakdown. (Reprinted with permission from Science Applications, Inc. and United States Bureau of Mines.)

In target area exploration, however, the uneven distribution of grade and abundance as well as the existence of microtopographic features give rise to different data needs. Stratified random sampling in a systematic pattern, usually in close grids and in large numbers is called for. Continuous profiles of the seafloor are required. The orientation of the ship track needs to be planned carefully.

The methods and equipment described above have been adequate in meeting needs in the prospecting stage. They are also useful for the target area exploration stage. Greater use of bottom sampling devices and longer runs of TV cameras allow a larger number of samples and wider areas to be covered. A better idea about the sediment properties helps refinement of the design of the collecting mechanism which has to be mobile over the sediments; in order to obtain requisite sediment data, bottom sampling devices that allow bringing up chunks of sediments, e.g. corers can be used to a greater extent. Sub-bottom profilers can also be useful for this purpose. Large amounts of nodules for the refinement of design of processing routes can be gathered by a greater use of drag dredges. Continuous profiles of the ocean floor areas showing gradients, and number and size of obstacles call for a more intensive use of TV cameras, narrow-beam up- and down-looking sonars, and side-scan sonars. There can also be a greater use of multiple instrument platforms. Nevertheless, there is a need for methods and equipment for obtaining more data at a higher level of confidence with less ship time. Faster deployment and recovery of bottom sampling devices and optical devices would reduce exploration costs. There is an acute need for continuous profiling of the seafloor with better resolution at less cost. Experiments with acoustic devices are being carried out to this end.

Chapter 5

Technology for Manganese Nodule Mining

Nodule mining technology developers had to address the basic question: how to pick up the nodules from the ocean floor and bring them up to the surface facility, most likely a ship. Three basic design concepts for mining technology have been pursued — picking up nodules with a dredge-type collector, and lifting them through a pipe; picking up nodules with a bucket-type collector and dragging up the bucket with a rope or cable; and picking up nodules with a dredge-type collector and having the collector ascend by the force of its own buoyancy.

Based on these three alternative concepts, three alternative mining systems have been developed or are being developed (Fig. 5):

> hydraulic mining system
> continuous line bucket (CLB) mining system
> modular or shuttle mining system.

HYDRAULIC MINING SYSTEM

This system uses the principles of hydraulics in lifting the nodules to the surface ship. A lift pipe, attached to the ship, extends close to the bottom of the ocean. A collector mechanism is linked to the end of the lift pipe. The collector picks up the nodules and feeds them into the pipe. The nodules are then pumped up through the pipe with hydraulic pumps fixed to the pipe; or they are sucked up through the pipe by means of compressed air injected into the pipe.

Collector Sub-system[12]

The collector is the most unique and complex sub-system of the hydraulic mining system. In the beginning, almost everyone

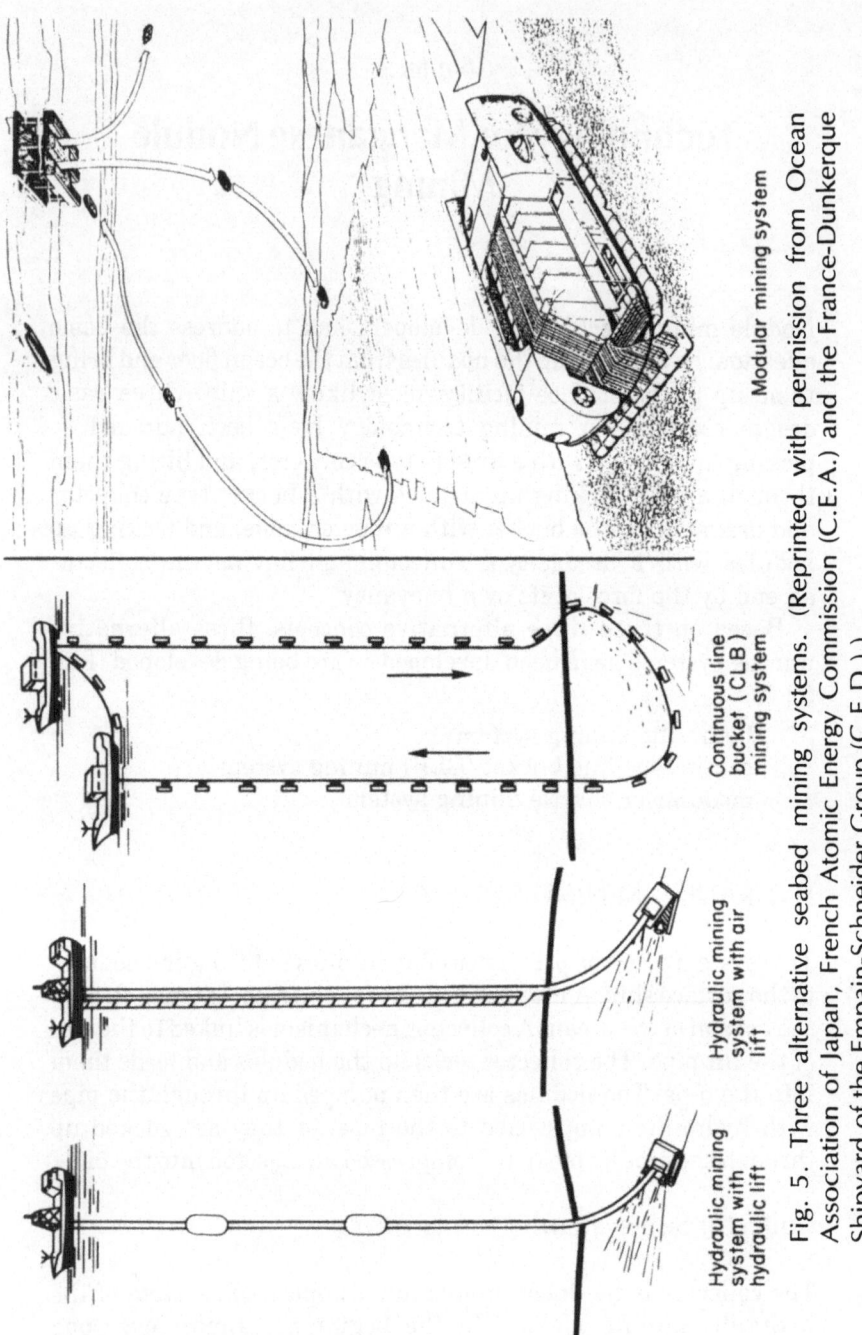

Fig. 5. Three alternative seabed mining systems. (Reprinted with permission from Ocean Association of Japan, French Atomic Energy Commission (C.E.A.) and the France-Dunkerque Shipyard of the Empain-Schneider Group (C.F.D.).)

Modular mining system

Continuous line bucket (CLB) mining system

Hydraulic mining system with air lift

Hydraulic mining system with hydraulic lift

envisioned a sort of simple dredge that would be towed and would suck nodules and sediments up a pipe to the mining ship. But, over the years, more than sixty patents have been issued on the collector and several alternatives have been developed and tested with various degrees of success.

The essential task of the collector sub-system is to collect nodules from the seabed, concentrate them and feed them into the vertical lift. In accomplishing this task, the sub-system has to be able to perform three groups of functions:

> collection and material processing,
> movement on the seabed,
> monitoring its own position and operation.

Specifically, the sub-system must have the capability to:

> dislodge nodules from the sediment,
> pick up nodules,
> concentrate nodules,
> wash away sediment,
> inject nodules into the lift sub-system,
> support itself and move on the seabed,
> manoeuver to control its direction of motion,
> locate and avoid obstacles in its path,
> monitor the operation of the sub-system,
> indicate its position to the surface ship.

Regarding the collection task, it should be recalled that nodules vary in size and shape. The collecting mechanism needs to be designed to retrieve an "average" nodule and must be able to discard objects that do not fall within a chosen size range. In view of the porosity and fragility of the nodules, the collecting mechanism must also be designed to have a sediment–nodule separation capability.

In respect of its material processing task, two basic design concepts have been developed. The first concept, which is referred to as a mechanical collector (Fig. 6), collects nodules by a mechanical method, crushes them, and injects the mixture into the lift sub-system. In the illustration shown in Fig. 6, two rows of teeth set at right angles to each other are used. One row deflects excessively large nodules or fragments from the collector's duct while the second row, consisting of closely spaced teeth, directs nodules of desirable size-ranges into the duct. The second concept, referred to as a hydraulic collector (Fig. 7), can be likened to a

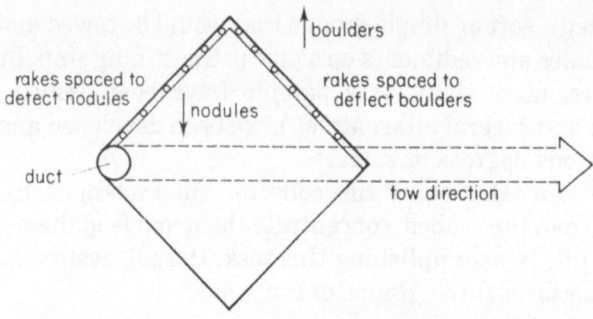

Fig. 6. Mechanical collector. (Reprinted with permission from Science Applications, Inc. and United States Bureau of Mines.)

sledge that moves over the seabed, pushing nodules into the sediment layer, and by means of a metal plate mounted beneath the sledge, forces the mixture of sediment and nodules up a raised duct. Jets of water to wash away the sediment are then injected into the duct, and the nodules that remain are then fed into the lift sub-system.

Regarding movement of the collector on the seafloor, account must be taken of the bearing, or shear, strength of the sediments.

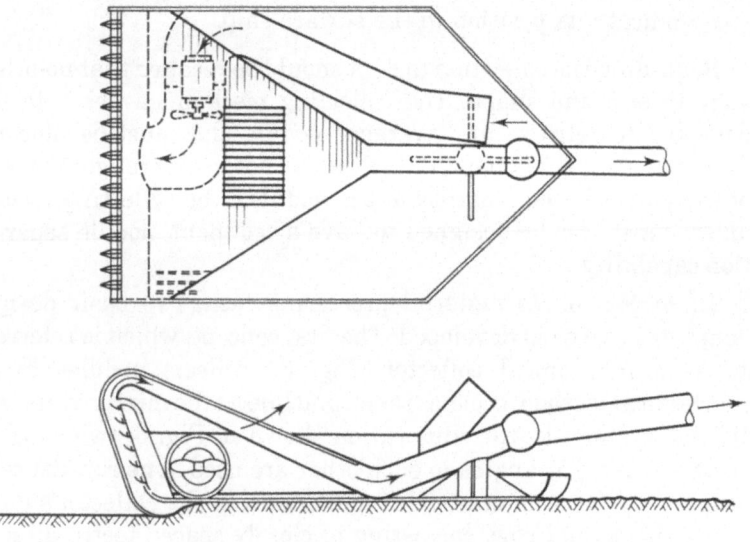

Fig. 7. Hydraulic collector. (Source: US Patent No. 3802740.)

The collector has also to be designed to handle some degree of slope as well as some of the smaller obstacles.

For movement on the seafloor, it has been proposed that the collector be either self-propelled or towed. One consortium, for example, favours a self-propelled remote controlled collector (Fig. 8). Using an Archimedean screw mechanism, the collector will be able to move along the bottom in all directions within the limits of its connexion with the lift pipe. Contact with the mining ship for remote control would be effected through a host of instrumentation placed on the collector. On the other hand, other consortia carried out tests with a towed collector. The towing member as envisioned would be the lift pipe. Some kind of traction mechanism, most probably skis, would be fitted to the collector. Some degree of steering could be achieved by using drag plates, pivots or thrusters.

For the monitoring task, various types of sensing devices are included in the collector sub-system. These types of equipment are used to: measure the rates at which nodules are collected and fed into the lift sub-system; show visual reports of the operation of the collector; provide acoustic information about the topography ahead and to the sides of the collector; measure pressure or depth, the attitude of the collector and its forward velocity; and provide information for position fixing of the collector relative to the mining ship and relative to previously mined tracks.

Lift Sub-system

The lift sub-system performs the basic task of lifting the nodules fed into the lift pipe from the collector to the surface ship. Two alternative lifting methods have been developed — hydraulic pump lift and air lift. In the former method, nodules are mixed with seawater to form a slurry and the slurry is pumped upwards with hydraulic pumps mounted in or on the lift pipe in a line at various depths. In the latter method, compressed air is injected into the lift pipe at various depths; the air–water mixture produces a density differential in the pipe. The mixture then moves upward under the influence of the hydrostatic head. This upward movement causes suction at the bottom of the pipe and the nodule slurry is sucked upward. Overall, the lift sub-system should have the capability to:

pump or suck up slurry,
control slurry flow,

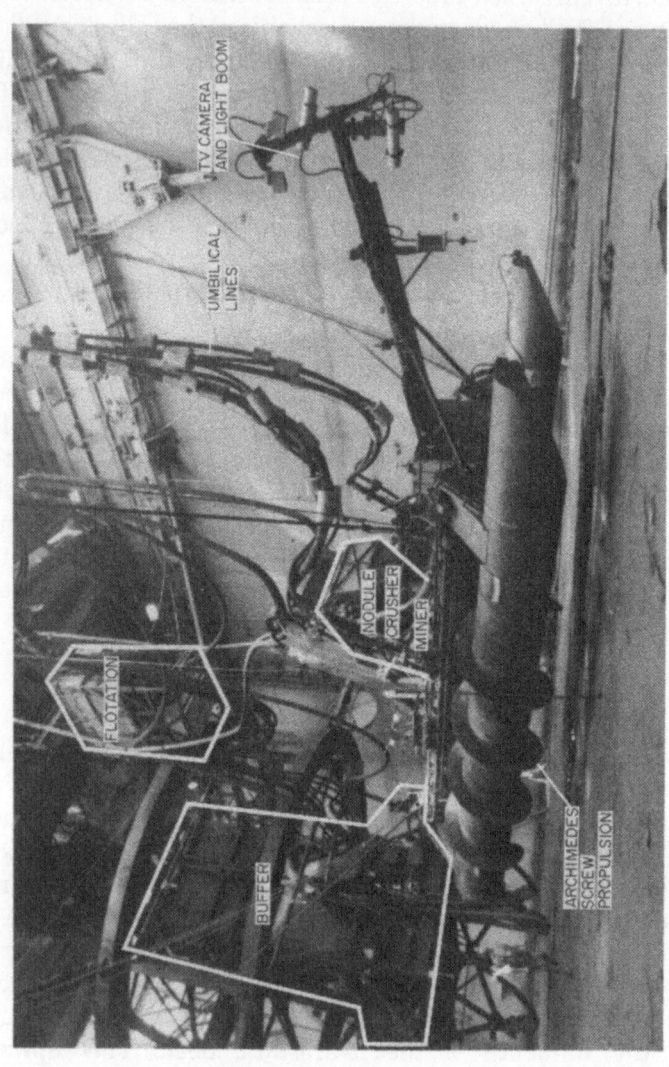

Fig. 8. Self-propelled collector. (Reprinted with permission from Ocean Minerals Company.)

work as a conduit of slurry,

provide a mechanical connection to the collector sub-system,

provide propulsion for the collector sub-system, if the collector is towed,

serve as a structural support for power cable and communication links to the collector sub-system,

withstand excitations caused by ship oscillations and movement through the water column,

withstand excitations of the pipe propagated through the collector sub-system encountering topographic variations,

avoid clogging in the pipe, particularly in case the slurry flow is shut down unexpectedly,

support its own weight along with the equipment and instrumentation attached to it.

For the lifting function, in the pump lift approach (Fig. 9), hydraulic pumps are mounted in series on the lift pipe at various depths to provide the required head. The position of the first set of pumps above the seafloor is dictated by cavitation considerations and is approximately at the depth of 1000–2000 m depending on the depth of the seafloor, and thus, the pump section generally occupies about 1000–2000 m of the upper pipe section. A two-phase flow involving nodules and water is carried out through the pipe. In the air lift approach (Fig. 10), air is injected into the lift pipe with the help of on-board air compressors. Air needs to be injected at various depths. Thus, an additional pipe parallel to the lift pipe is required to carry the compressed air. The length of this pipe is approximately 1000–2000 m. A three-phase flow of air, water and nodules is thus entrained in the lift pipe.

In view of the porosity and fragility of the nodules the pumping function has to take into account two factors: slippage associated with the flow of the water–slurry mixture in the pipe and friction associated with the impacts among solid particles and between solid particles and the pipe wall. Pipe pressure gradients and pump power requirements are determined by the constraint that the effects of these two factors are kept at an acceptable level and that sufficient power is provided to lift the nodules, sediments and water from the seafloor to the sea surface.

The diameter of the pipe will be geared for the optimum transport of slurry. However, variations in nodule abundance make it necessary to provide buffer storage facilities in case more than the optimum amount of nodules is collected.

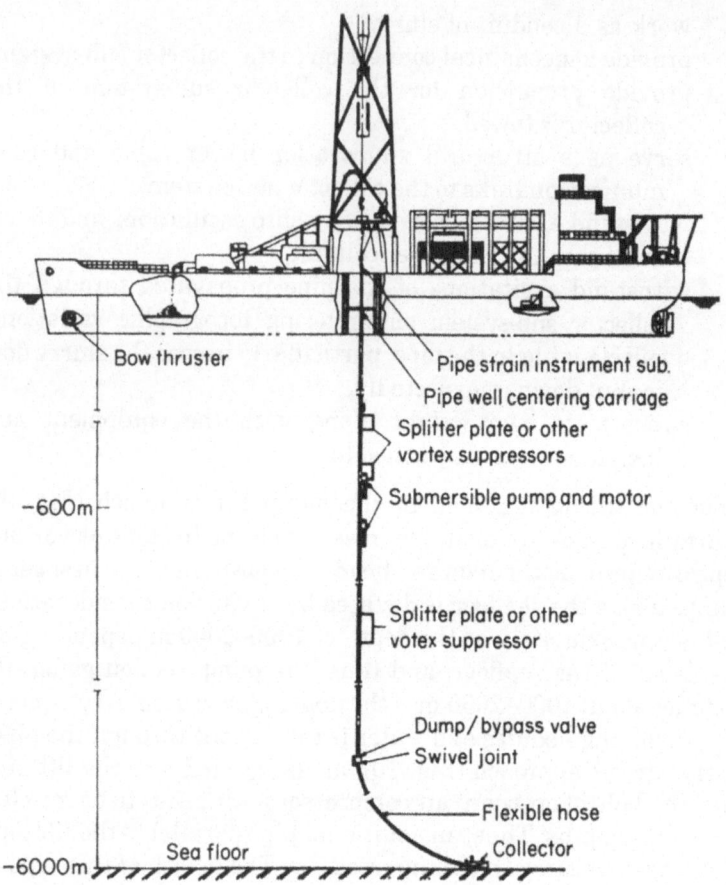

Fig. 9. Hydraulic pump lift. (Courtesy of Ocean Management, Incorporated.)

The lift sub-system functioning as a pipe string needs to take into consideration the effects of several forces. The motion of the mining ship will induce a load on the pipe string; towing the pipe by the ship through the water column will result in a drag on the pipe string: vortices will be shed from the sides of the pipe that could excite pipe vibrations beyond desirable levels. Several measures can be taken to minimize the adverse effects of these forces. High strength steel can be used for fabricating the pipe so that it can withstand some of the effects. The speed of the mining ship and the diameter of the lift pipe can be determined in a manner that drag is limited to an acceptable level. It is expected

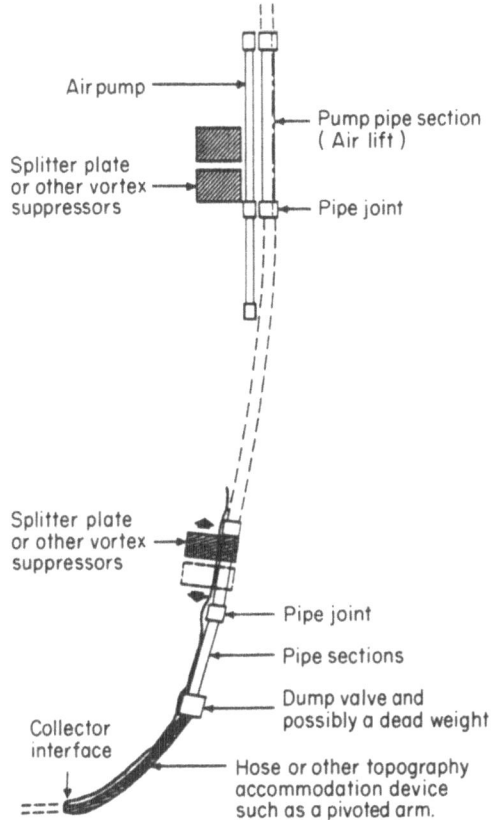

Fig. 10. Air lift. (Reprinted with permission from Science Applications, Inc. and United States Bureau of Mines.)

that the diameter of the pipe could be in the range of 30 to 75 cm and the speed of the ship would be limited to one to three knots. The mechanisms to reduce the effects of ship's oscillations are discussed later. Vortices can be suppressed through the selective use of splitter plates or fairings.

Finally, a valve at a proper location may be used to avoid clogging in the pipe, in case the vertical flow of the slurry shuts down unexpectedly and solid material begins to accumulate at the mouth of the pipe.

The link between the pipe string and the collector sub-system has to withstand bending without significant stresses, because it is subject to high curvature. The link also has to provide room for accommodating variations in local water depths and bottom topography. These requirements can be achieved by the use of a pivoted truss, articulated pipe sections or a properly supported flexible hose, the last being the most likely.

Mining Ship Sub-system

The essential function of the mining ship sub-system is to receive nodules from the lift sub-system and transfer them to the transport system. The mining ship sub-system has to have the capability to:

provide structural support for the subsurface sub-systems — the collector and the lift sub-systems,

provide means to assemble, deploy, operate, monitor and recover the subsurface sub-systems,

supply power to the subsurface sub-systems,

propel the whole mining system over the mine site, possibly according to a pre-determined mining plan,

provide for transfer of ore to the transport system,

provide buffer storage for accumulated ore and also storage for subsurface sub-systems when not in use,

control the whole mining and ore transfer operation,

serve as a hotel, storehouse and repair shop.

The function of handling the subsurface sub-systems — lift and collector — requires that the mining ship gives protection to these sub-systems from the motion and load effects of its own oscillations. There are six types of oscillations of a ship — surge (linear motion forward and backward), sway (linear motion from side to side), heave (linear motion up and down), pitch (rotation of the bow and the stern around the horizontal axis), roll (rotation of the sides around the side-to-side axis) and yaw (rotation around a vertical axis). Heave tends to make the subsurface sub-systems move up and down and pitch and roll tends to induce angular motion in the lift sub-system having a bending effect. To minimize the up-and-down movement of the lift sub-system, a heave motion compensator on the ship can be used which supports the ship–pipe interface with hydraulic cylinders holding the interface at a constant height (within a certain limit) while the ship undergoes motions in waves. To reduce the effects of pitch and roll on the pipe bending stress, a gimbal can be useful.

For the deployment of the subsurface sub-systems, a moonpool is required in the mining ship. The pipe string is suspended from the pipe support equipment through the moonpool of the ship.

The propulsion and navigation functions first require positioning the mining ship *vis-à-vis* the collector. Ship positioning can be accomplished by several methods: positioning at one- to two-hour intervals by the satellite system; positioning by acoustic measurements by utilzing acoustic transponders placed on the seabed, a method known as long baseline acoustic positioning; and continuous three-dimensional positioning by a new system to be operational in the mid-1980's, called the Global Positioning System. In the first method, positions during the intervals can be determined by the use of moored buoys. This method can determine a ship's position within 60 m, which may not offer sufficient accuracy for nodule mining operations. In the second method acoustic transponders placed on the seabed can provide positioning with accuracy of 5 to 10 m, but this method requires placing transponders at short intervals. A modification of this method is being developed which will combine transponders with moored buoys. Once the position of the ship is known, a computer control system can determine required engine corrections to cause the ship to follow the desired path and can automatically operate appropriate thrusters.

Because of the variation in grade, abundance and seafloor characteristics, the configuration of the mineable areas may be quite complex. The mining ship must have manoeuverability to cover the mineable areas without spending too much ship time as well as avoiding risks of the collector colliding with large obstacles. The ship also has to have sufficient manoeuverability to take advantage of favourable weather heading. It has to be capable of weathering a high degree of roughness in sea states. Over the span of a year, it is likely that the mining ship will encounter sea states rougher than it is prepared for. Contingency plans for retrieving the subsurface sub-systems and moving to fairer weather sea surface have to be made.

Methods of separating ore from water need to be available either in the mining ship or the transport ship. Ore storage capacity also needs to be incorporated in the mining ship. This means that the size of the mining ship will be much larger than that of the existing deep water drillships. The hull of the mining ship, however, will be similar to that of a deep water drillship.

Figure 11 presents a breakdown of the components of the hydraulic mining system.

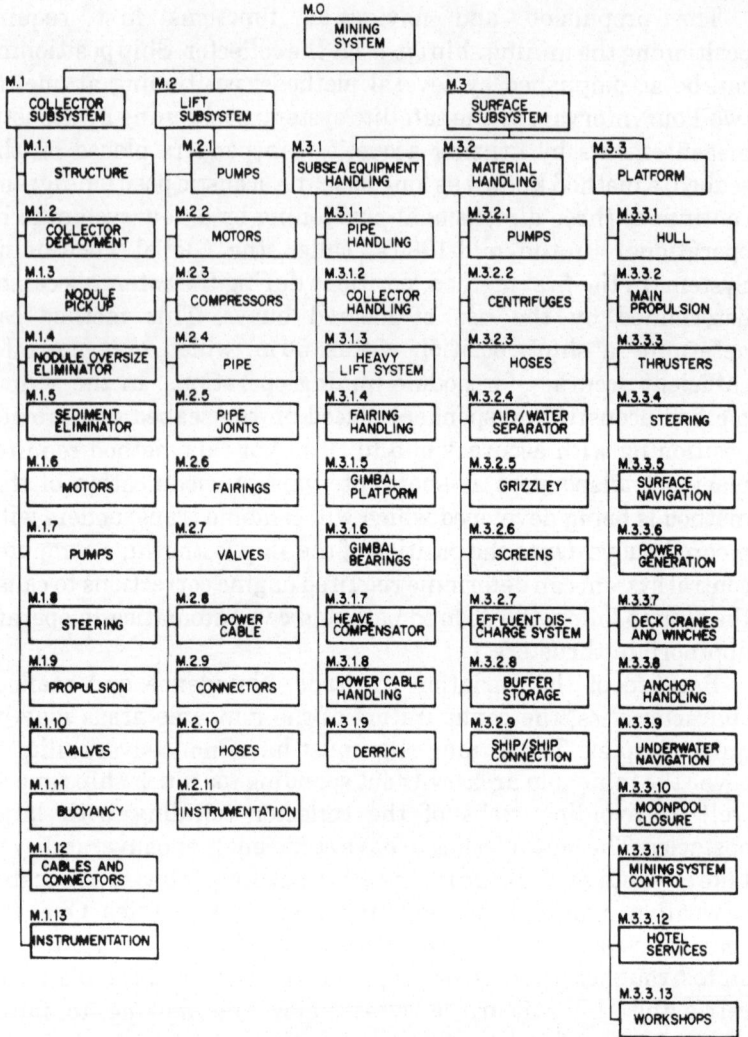

Fig. 11. Mining technology component breakdown: Hydraulic mining system. (Reprinted with permission from Science Applications, Inc. and United States Bureau of Mines.)

CONTINUOUS LINE BUCKET (CLB) MINING SYSTEM

The CLB system can be either a one-ship or a two-ship system. The original CLB system used one ship, empty buckets going down from the stern and partially filled buckets coming in at the bow.

The distance between the downward-moving and the upward-moving parts of the loop of the rope is dependent on the length of the ship and thus cannot be altered. The curve of the loop is divided into four parts: descending part, waiting part, dredging part and ascending part (Fig. 12). This curve of the loop is determined by the length of the cable and the speed of the ship. Entanglement can be avoided by achieving an optimum combination of these two factors and the use of hydrodynamic deflectors. But the optimum combination may be disturbed by various other factors — obstacles on the seafloor, underwater currents, variations in the ship's course because of weather heading, bad weather conditions, etc.

To minimize these disadvantaes and to add to the flexibility of operation, a two-ship system has been designed. The empty buckets would go down to the ocean floor from one ship and would be dragged up in another sister ship. The ship dragging up the buckets may be in the back or the front of the other ship. The distance between the descending and ascending parts of the loop and the curve of the loop can be influenced by the relative positioning of the two ships. In fact, the length and form of the part of the loop on the seabed carrying out the nodule collection operation can also be varied, either to have a longer portion with more buckets on the seabed to increase pick-up rate or to lift up the loop from the seabed to avoid obstacles. All this flexibility could be achieved by varying the speed of the whole two-ship system and

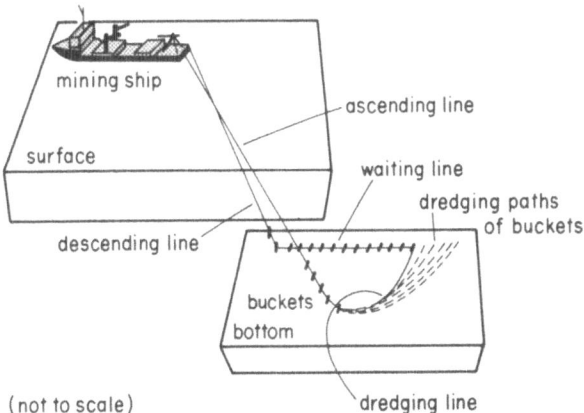

Fig. 12. The loop in the CLB mining system. (Reprinted with permission from K. Handa and N. Yamakado.)

the relative distance between the two ships, the positioning of the two ships and the speed of the circulation of the loop. The two-ship system, however, would require higher navigational and positioning capabilities.

Rope Sub-system

In the CLB mining system, buckets have to pass from stern to bow — attached at stern decks, removed at bow deck and filled buckets have to be unloaded on deck. Both the bucket passing and the unloading mechanisms need to be smooth. Smooth circulation of the rope with many buckets attached to it require traction devices on the ship. Umbrella type wheels can be used as traction devices.

The problem of tangling of the rope may arise while the buckets are moving as well as when the movement is stopped. In the former case, appropriate traction devices for smooth circulation is helpful. In the latter case, hydrodynamic separators can be used.

The type of rope used in the CLB system is very important not only because of its carrying strength but also because of its vulnerability to entanglement. It was found that compared to any other type of rope, braided rope made of polypropylene reduces the risks of entanglement most.

Special types of buckets can be used and attachment of the buckets to the rope can be done in such a way as to fix four points of a bucket to the rope so that a force pressing the buckets outward during dragging can be generated. This reduces the risks of entanglement.

Bucket Sub-system

The buckets have to be able to dislodge nodules from sediments. This can be done with the help of teeth attached to the buckets, but teeth may have a tendency to get embedded in the sediment. It was found that the use of cutting edges with sleds under the buckets so that the buckets just graze the surface is more effective. The buckets are most likely to have nets attached to them in order for the sediments to be washed out. But depending on the mesh size of the nets, very small-sized nodules may drop through the nets.

Figure 13 gives a breakdown of the components of the CLB mining system.

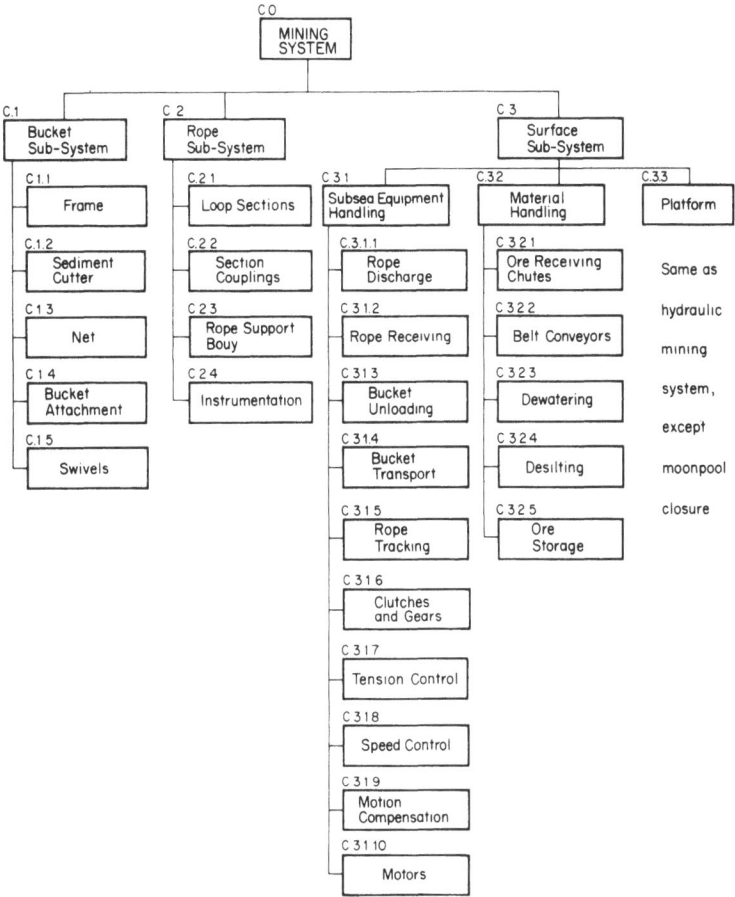

Fig. 13. Mining technology component breakdown: CLB mining system.

MODULAR MINING SYSTEM

In this system, the collector is launched with ballast material such that the weight in water of the ballast is equal to the weight in water of the nodules to be collected. The collector is designed to have sufficient buoyancy so that the vehicle is weightless in water. Thus, in descent, thrusters propel the unit down steadily against hydrodynamic resistance alone. On the bottom, the collector is propelled over the bottom, and as collection proceeds, ballast material is simultaneously ejected on an equal weight in water basis. In this manner, a small net weight in water of the collector is maintained. Mining is terminated shortly before ballast material ejection. Ballast material ejection is continued until the weight of the vehicle is zero or slightly negative. Finally, the vehicle is propelled by thrusters to the surface, docked with the surface ship, unloaded, serviced, and re-ballasted for a new mining cycle. In theory, very little onboard power is required to collect the nodules — the major source of energy is the potential energy of the ballast material. The operating principle of this system is illustrated in Fig. 14.

There are many ways in which a modular mining system could be designed. However, generally several functions need to be performed. These are outlined briefly below:

Collector (or Shuttle) Unit

descend to a prescribed point on the bottom by remote commands or inertial control;

mine and lift in accordance with remote commands or onboard programmes

dislodge and pick-up nodules,

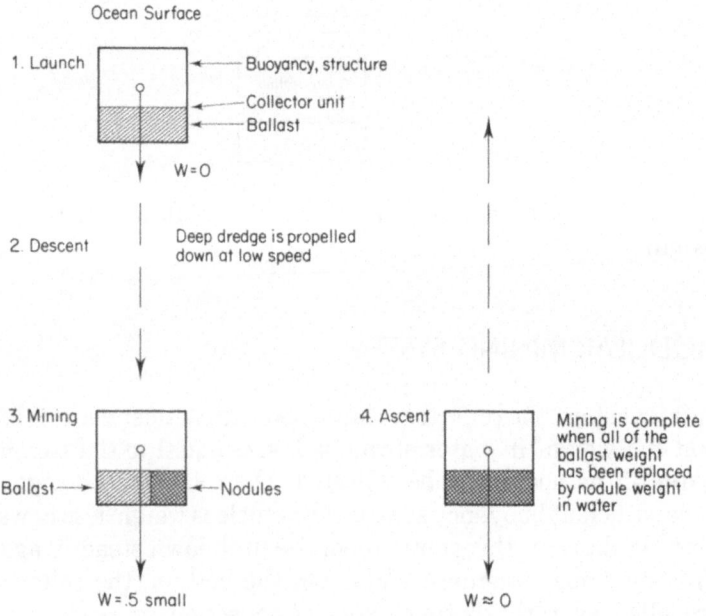

Fig. 14. Operating principle of the modular mining system.

transfer nodules to onboard storage while ejecting ballast
material at an equal weight in water rate and so that
the weight balance of the collector is maintained,

provide onboard storage for ballast material and nodule
payload,

provide propulsion to traverse the bottom during mining
and ascent/descent,

provide buoyancy such that the weight of the vehicle in
water with the payload is essentially zero,

provide onboard energy storage for propulsion, collection,
and load levelling,

provide an emergency location system to facilitate location
and recovery of stalled or strayed units,

provide sensors and controls as necessary to support the
preceding functions.

Surface Platform

provide launch and docking interface — most likely, a moonpool,

provide system control commands through remote commands
(acoustic) or through programmes stored onboard the
units,

unload recovered units,

inspect recovered units,

perform scheduled and unscheduled maintenance,

load ballast material and balance the unit prior to launch,

provide recovery system for stalled or strayed units.

Recently, Association Française pour l'Etude et la Recherche
des Nodules (AFERNOD) documented one approach to modular
mining. This approach would utilize several remotely controlled
shuttles as collectors. Each shuttle would make about 3 trips/day
and would have a production capacity of 250 tons per trip. The
shuttle units would be controlled by an acoustic link from the
surface platform. Figure 15 illustrates the major components of
the collector. These are identified below the figure. The overall
dimensions of the collector would be 24 m × 12 m × 7.5 m and the
collector would have an empty mass of 550 tons.

(a) *Nodule/Ballast Material Storage Containers.* These are
required to house the ballast material and nodules.

(b) *Propulsion During Mining.* For propelling the collector,
four independent Archimedean screw thrusters would be
used. These must be controlled independently because load

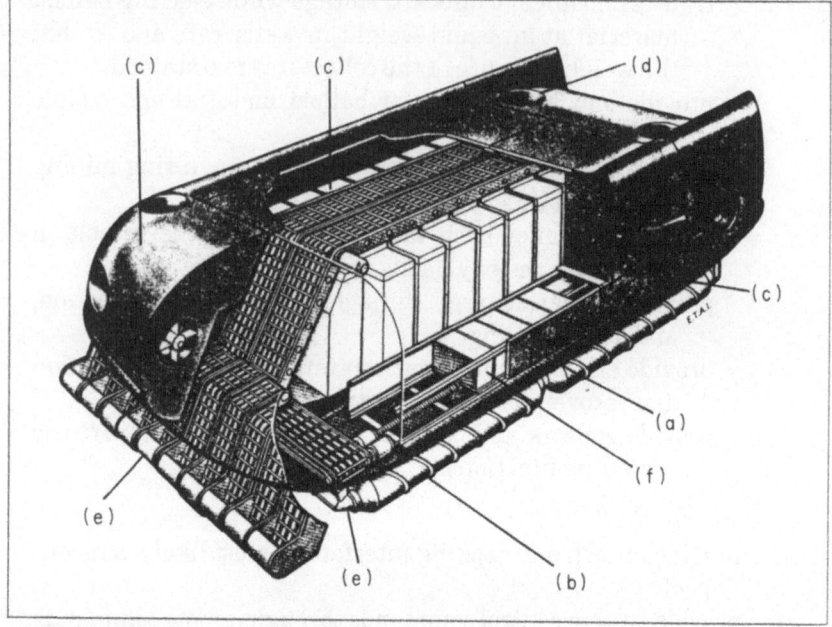

Fig. 15. Proposed collector (shuttle) of AFERNOD. (Reprinted with permission from French Atomic Energy Commission (C.E.A.) and the France-Dunkerque Shipyard of the Empain-Schneider Group (C.F.D.).)

KEY: (a) nodule/ballast storage containers; (b) Archimedean screw thrusters for bottom propulsion; (c) syntactic foam for buoyancy; (d) thrusters for ascent and descent; (e) mechanical collection device, longitudinal and lateral conveyors for nodule transfer; (f) removable batteries.

levelling errors and variations in sediment properties result in yaw torques on the vehicle. The overall net weight of the vehicle must be maintained within defined limits so that optimum sediment penetration for movement would occur without encountering undesirable sediment resistance.

(c) *Buoyancy.* Syntactic foam, which is a standard buoyancy material for deep sea submersibles, would be used.

(d) *Propulsion During Ascent/Descent and Docking.* Propulsion assistance during ascent and descent would be provided by thrusters to control the altitude and trajectory of the collector. The thrusters would provide pitch, roll and yaw control of the unit as well as propulsion for trajectory control in ascent, descent, and rendezvous. The thrusters would also compensate for any hydrodynamic imbalances resulting from load levelling errors throughout the ascent/descent phases.

(e) *Collection and Ore Distribution.* To dislodge, collect and transfer nodules to the storage containers, a collection device and longitudinal and lateral transfer conveyors would be used.

(f) *Batteries.* Power would be provided by removable batteries located on each side of the vehicle.

The AFERNOD approach would utilize a large, 100 m × 100 m semi-submersible surface platform with 140,000 ton displacement capacity and 56 m operating draft. The collector units would dock, and would be relaunched, from 4 underwater ports located 40 m below the water line. Ore off-loading, ballast loading, and battery replacement operations would be performed under water. The platform would include storage for about 60,000 tons of nodules or ballast material and accommodation for 150 to 180 people.

MINING PLAN

Regardless of the system chosen for a mining operation, careful plans must be made to attain desired production goals in a given time span. Practice established at many land operations will probably be followed — an ore reserve map of a selected mine site will be made, and superimposed on it will be the mining plan and a layout of areas expected to be mined in successive years. The map will show the mineable areas above the selected cut-off grade and abundance and with the acceptable seafloor characteristics, low grade and low abundance areas to be bypassed and areas with obstacles and other seafloor hazards to be avoided. This seafloor map will identify a number of blocks within the mine site and the nodule tonnage and metal tonnages expected to be mined from different blocks. The direction of the collector and its sweep pattern will be indicated.

The overall mining plan will consider, among other things, manpower requirements, supplies of spare parts, and scheduling with regard to equipment maintenance, and ore transfer. Contingency plans will be made for accidents or other interruptions in operations, for the deployment and recovery of submerged equipment, and for lift failure, and collector and other mechanical breakdowns. Procedures will be established for periods of adverse weather and/or transfer operations. Also, logistical problems such as shift rotation, fuel, food and equipment provisioning, and crew amenities will have to be worked out. Good communication will have to be maintained with shore bases, process plants, ore carriers, and weather stations. In general, the mining plan is extremely important to the operation efficiency of a remote, isolated activity that is vulnerable to capricious weather.

Chapter 6

Assessment of Nodule Mining Technology

TESTS OF NODULE MINING TECHNOLOGY

All three mining systems described in Chapter 5 are possible candidates for commercial application by first-generation seabed miners; laboratory studies and/or small-scale at-sea tests have led the technology developers to conclude that the systems are basically feasible. The CLB system was the first to be tested in deep water. There have been several at-sea tests of the hydraulic mining system during the last few years. The modular mining system is still at the design stage; it is yet to be tested.

Of the four multi-national consortia, Ocean Mining Associates (OMA) tested an air lift system with a towed collector during 1977 and 1978. Approximately 500 tons of material were recovered with a system with a design capacity of 1200 tons per day. Ocean Management, Inc. (OMI) conducted tests with both hydraulic pump and air lift and towed collectors in early 1978 and recovered 1000 tons over a few days. Ocean Minerals Company (OMCO) tested an air lift system with a remotely controlled self-propelled collector in 1978 and 1979. The Kennecott group (KCON) has not conducted a mining system test although they tested the capabilities of a towed collector in late 1974 and early 1975.

The French group, Association Française pour l'Etude et la Recherche des Nodules (AFERNOD), and the Japanese group, recently re-named Deep Ocean Resources Development (DORD), appear to have continuing interest in the CLB mining system, especially a two-ship system. AFERNOD participated in the tests of this system in 1970 and 1972 and has done development work. It is also considering the development of a modular mining system. DORD is yet to conduct a mining system test. Not much information is available about the mining system or sub-system development work of the three publicly sponsored entities from the USSR, India and China.

57

However, the studies and/or the tests also pointed to some engineering problems in each system that need to be worked out. Improvements, or changes, in design are required in order to achieve better system performance. Also, all the above at-sea tests were conducted with small scale experimental or pilot systems for short lengths of time; a commercial scale venture however, requires design and operation of a much larger (perhaps by an order of magnitude) mining system. The transition from small scale to commercial scale will involve further engineering considerations.

PERFORMANCE OF NODULE MINING TECHNOLOGY

At this point, only very rough indications of the performance of technology is possible. Any reliable estimate has to wait until the large-scale prototype operation is carried out, the results of which will have to be reasonably successful before the developers decide to proceed to a commercial operation. However, some important observations can be made based on the current status of technology, as determined from publicly available information. This section will attempt to analyse mining technology in terms of its performance. The discussion will necessarily focus on the hydraulic mining system in view of relative scarcity of information about the other two systems.

Two aspects of performance are considered — reliability and efficiency. Reliability involves availability of the mining system for operation. One of the measures of reliability is mean-time-between-failures (MTBF) expressed as the average number of days between serious interruptions of operation due to system or sub-system failures. An MTBF of 45 days or less is considered unacceptable.

Collector Efficiency in the Hydraulic Mining System

The efficiency of the mining system is dependent predominantly on the efficiency of the collector sub-system. It will be worthwhile to examine what is involved in the efficiency of the collector sub-system.

As the collector passes over the seafloor, it covers a particular track in one passage. This is the collector swath. For various reasons, the collection mechanism may not act upon the whole swath. For example, if a collector has a running gear, the running gear itself may take up a portion of the swath and may preclude collection of nodules in that portion. Part of the swath on which the collection mechanism acts expressed as a percentage of the whole swath is called swath efficiency (designated by e_s).

The collector may not pick up all the nodules in the part of the swath on which the collection mechanism acts. Very large nodules or very small nodules may be rejected, depending upon the collector design. If the collector is preceded by blades, tines or jets, they may bury or make unavailable a portion of the nodules. The quantity of nodules that is picked up by the collector as a proportion of the quantity of nodules presented to the collector is called pick-up efficiency (designated by e_p).

Taken together, swath efficiency multiplied by pick-up efficiency ($e_s \times e_p$) determine the quantity of nodules picked up from the swath.

After nodules are picked up, they are appropriately concentrated to be fed into the pipe. In the process of concentration, some losses can occur. This is because in some collector designs, excess water and sediment are separated and removed before the concentrated nodules are fed into the pipe. The quantity of nodules that is fed into the pipe as a proportion of the quantity of nodules picked up is called concentration efficiency (designated by e_c).

Swath efficiency, pick-up efficiency and concentration efficiency, combined together ($e_s \times e_p \times e_c$), determine the quantity of nodules from a swath that is fed into the pipe.

Once the nodules are fed into the pipe, even under the assumption that no losses occur in the process of lifting through the pipe, losses associated with the separation of nodules on the surface facility and their transfer to the transport vessel should be considered. The quantity of nodules transferred to the transport vessel as a proportion of the quantity fed into the pipe is called transfer efficiency (designated by e_t.)

Taken together, swath efficiency, pick-up efficiency, concentration efficiency and transfer efficiency ($e_s \times e_p \times e_c \times e_t$) determine the quantity of nodules transferred to the transport ship *from a given swath* in the course of one passage. This is usually referred to as dredge efficiency, e_d.

In subsequent passages of the collector, there may be overlaps with previously mined swaths. If the overlap as a proportion of the swaths is designated by o, then the non-overlap part as a proportion of the swaths is called feed efficiency (designated by $e_f = 1 - o$). (See Fig. 16.)

Feed efficiency combined with the four efficiencies above $((e_s \times e_p \times e_c \times e_t)\, e_f)$ determine the quantity of nodules transferred to the transport ship *from a particular track* on the seafloor. This is called the efficiency of the collector sub-system (designated by e_{cs}.)

The width of the swath (in metres) designated by S, gives the size of that particular track. The velocity of the collector (in metres per second), designated by V, combined with the width of the swath $(S \times V)$ give the area of the seafloor covered per second. If average abundance of nodules is designated by A (wet kg per square metre) then $S \times V \times A$ gives the rate at which nodules are encountered by the collector sub-system.

The collector sub-system efficiency, combined with this rate $(e_{cs} \times S \times V \times A)$, give the yield of the sub-system (wet kg per second), designated by Y. The annual production rate, P, in metric tons, can be obtained from this rate by converting kg's to metric tons and seconds to days (K is a constant representing this conversion)

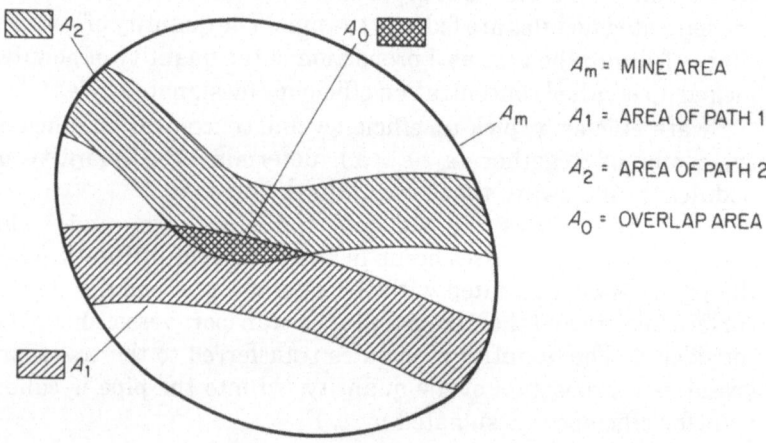

A_m = MINE AREA

A_1 = AREA OF PATH 1

A_2 = AREA OF PATH 2

A_0 = OVERLAP AREA

Fig. 16. Areal efficiency and feed efficiency. (Reprinted with permission from Science Applications, Inc. and United States Bureau of Mines.)

Areal efficiency, $\quad e_a = \dfrac{A_1 + A_2 - A_o}{A_m}$

Feed efficiency, $\quad e_f = \dfrac{A_1 + A_2 - A_o}{A_1 + A_2}$

and multiplying by D, number of days for which the mining system is available for production, i.e. $P = Y \times K \times D = e_{cs} \times S \times V \times A \times K \times D = e_d \times e_f \times S \times V \times A \times K \times D$.

While the preceding discussion concentrated on the efficiency of the mining technology in terms of obtaining nodules from a particular mined area, another way of looking at the efficiency is to consider what part of mineable areas are actually mined by the application of the mining technology.

The collector is unlikely to cover each and every metre of the mineable areas, because of the lack of precision of collector or ship manoeuverability. On the other hand, for the same reasons, a collector may pass through the same space more than once. The proportion of mineable areas mined at least once is expressed as sweep efficiency or areal efficiency, e_a. (See Fig. 16.)

Table 4 presents current estimates of the various efficiencies defined above. At this point, technologies for nodule mining are unproven and obviously, technical uncertainties about their reliability and efficiency persist. In fact, even concerning cost-effectiveness of technology (costs associated with the application of technology being lower than the revenues earned by its application), any conclusions can at best be conditional. However, it will be extremely useful to study the possible ways improvements can be made in technology leading to higher reliability and efficiency. It is important to bear in mind that in an interrelated system, any change in one part will have implications for the overall configuration of the system. Changes in overall configuration may be required to an extent that the system design concept itself may have to be modified drastically.

Possible Improvement in Collector Efficiency

It is evident from the relationship between the annual production rate and the parameters characterizing mining technology, i.e. $P = e_d \times e_f \times S \times V \times A \times K \times D$, that the annual production rate can be increased by:

> increasing dredge efficiency, e_d
> increasing feed efficiency, e_f
> increasing swath width, S
> increasing collector velocity, V
> increasing average abundance of nodules, A
> increasing number of days the system is available for operation, D
> a combination of the above.

Table 4. Estimates of Various Efficiencies of the Collector Sub-System; Relationship with System Design

Name		Description	Estimate	Relationship with System Design
Swath efficiency	e_s	Fraction of the collector swath on which the collection device acts	0.8—0.9	Depends on running gear design, i.e. running gear may bury nodules before they can be picked up.
Pick-up efficiency	e_p	Fraction of nodules on which collection device acts which are actually collected	0.9—1.0	A number of systems achieve high pick-up efficiencies. Associated power requirements and reliability are key issues. Sensitivity to speed is also important.
Collector concentration efficiency	e_c	Fraction of collected nodules injected into lift system	0.8—0.9	A separator may be provided in the collector at the bottom of the lift pipe. The percentage of nodule material lost can probably be more effectively reduced by locating the separator on the lift pipe because this reduces the mass and weight of the collector.

Efficiency	Symbol	Definition	Value	Comments
Surface separation and transfer efficiency	e_t	Fraction of nodules received on the surface which are transferred to the transport ship	0.9	Some nodule fines may be lost in shipboard dewatering and transfer. This may be reduced by improving dewatering capabilities.
System or dredge efficiency	e_d	Fraction of nodules in collector swath delivered to transport ship		Depends on other system efficiencies discussed above. ($e_d = e_s \times e_p \times e_c \times e_t$)
Feed efficiency	e_f	Fraction of the collector swath which was not previously mined	0.8–0.95	Any feasible system must achieve a high feed efficiency to allow reasonable size and operating speeds. Steered and powered collectors are projected to achieve a given feed efficiency simultaneously with a high areal efficiency.
Areal efficiency	e_a	Percentage of mineable area swept at least once by the collector	0.45–0.95	Towed unsteered collectors may achieve values near the lower end of this range, 0.45–0.6. Powered and steered collectors may approach areal efficiencies of 0.95.

Out of the four components of dredge efficiency, there may be room for improvement only in pick-up efficiency and concentration efficiency. Pick-up efficiency can be increased by increasing power of the collector, either transmitted from the ship or installed in the collector itself. Increasing concentration efficiency may require additional equipment to raise concentration capacity and/or to accommodate higher flow rate. Increased dredge efficiency will mean higher input rate to the lift pipe and this will require increasing the diameter of the lift pipe and, thus, power needed for lifting and capacity for pipe handling.

Feed efficiency can be increased essentially by reducing the possibilities of overlap with a previously mined swath. This can be done by increasing collector steering accuracy or by increasing swath spacing.

Feed efficiency can be increased by spacing swaths in a manner as to have "safe" distances between swaths. This means spaces are left unmined adjoining mined swaths. Thus, increasing production rate through increased swath spacing necessarily requires a larger and/or a greater number of mineable areas and a larger mine site.

MAJOR SUB-SYSTEM OPTIONS IN THE HYDRAULIC MINING SYSTEM

There are several key design options in the sub-systems of the Hydraulic Mining System. Among these are:

steered versus unsteered collectors
air lift versus hydraulic pump lift

There is a considerable debate among the major consortia about the selection between the alternatives. It is possible for two equally competent design teams to arrive at alternative choices based on differences in basic assumptions, judgements or risk aversion preferences. Also, each design team is influenced by its technical experience and innovativeness as well as its relationships with outside suppliers. Site-specific physical factors are also key elements in the selection between alternative designs. Some consortia have carried out experiments with both the alternatives and have yet to settle on one. Ultimately, the choice among the alternatives may not become apparent until after large-scale prototype operations for a length of time at a given site.

Steered versus Unsteered Collector

Collector steering may be provided to perform the function of obstacle avoidance and/or to increase areal efficiency while maintaining a required feed efficiency. A capability to perform the first function increases system reliability, and thus, is important in assuring achievement of production rate goals. A capability to perform the second function will tend to maximize yield from given mineable areas.

The most critical criteria for a decision between the two approaches are their respective potentials for achieving a designated daily production rate and reliability. The steered collector is expected by some to be capable of a higher areal efficiency than the towed collector. Others think that the advantages of the steered collector would be offset by its greater complexity, which possibly could lead to higher frequency of breakdown. They also think that an unsteered collector which is a simple and rugged collector is more likely to achieve high reliability. Of course, they could both be correct depending on the physical characteristics of mine site.

Two points should be kept in mind in this connexion: (1) since neither design has been tested for large scale production, the differences in areal efficiency and reliability can only be considered potential; and (2) implications for unnecessary waste of resources need to be considered.

Steering forces can be provided to the collector through propulsion mechanisms. The collector can be self-propelled, as mentioned in Chapter 5. It should be noted that in the hydraulic mining system, the collector remains linked to the lift pipe whether self-propelled or not. Self-propelled collectors are relatively free from the effects of ship motion as well as of pipe dynamics. Self-propelled collectors can be remotely controlled from the mining ship. Power requirements for this type of collector are higher than a collector propelled by the towing member, i.e. the pipe only. However, ship steering requirements are reduced if steering is incorporated on the collector, thus reducing energy consumption in ship manoeuvres. It is to be noted that collectors which are not can achieve some degree of steering through drag plates, pivoted skis or thrusters; however, these do not provide the precision of steering achieved with a remotely controlled self-propelled collector.

Air Lift versus Hydraulic Pumps

The choice of lift design has been a major subject of investigation by the technical teams of the various consortia. The decision between two of the leading designs, the air-lift and the hydraulic pump lift, is not readily apparent from simple engineering investigations. In fact, some consortia are planning tests of both lift approaches in their large-scale demonstrations. The two designs are compared in terms of their advantages and disadvantages in Table 5.

MINING SYSTEM OPTIONS

Each of the three mining systems described in Chapter 5 offers various advantages and entails various disadvantages, relative to the others. Each system has key technical issues which are the focus for future development efforts. Depending on the success of these efforts, their operational performance and thereby the

Table 5. Comparison of Air Lift (AL) and Hydraulic Pump Lift (HP)

Criterion	Comparison
Costs	Capital costs are about the same for the two designs. For the same production rate, energy costs are higher for AL, but HP costs are higher in terms of wear and maintenance. Maintenance costs are higher for HP because the sub-system must be partially recovered to maintain the sub-system.
Reliability/ availability	The designs should be about equally reliable (equal mean-times-between-failures), however, the consequences of a failure are more severe for HP in that significant down-time is incurred in recovering the sub-system. The AL sub-system has no moving parts below surface. Air compressors are located on the ship where they can be readily maintained. Air lines may be subject to leakage, however.
Handling	The AL could present significantly greater handling problems in that it is essentially a two-pipe sub-system. In addition, the fairings required are much larger than for the HP sub-system.

cost-effectiveness of their application in an actual commercial-scale mining venture, will be determined.

Most development work to date has been on the hydraulic system — carried out by four different consortia. If this trend continues it will indicate that the consortia perceive a technical advantage in the construction and operation of the system — at least for first generation operations.

Hydraulic Mining System (HMS)

The HMS offers far more flexibility for control of the collector sub-system than does the CLB. With a powered and steered collector, such as a remotely controlled self-propelled collector, the system can achieve high areal efficiencies and avoid obstacles down to a small size. The system can be designed as sufficiently rugged to surmount obstacles of a size which is readily detected by previous survey. The technologies for drillship, slurry hydraulics, and dredging are perhaps more readily extended to the HMS than are related technologies extendable to the CLB, or MMS.

The principal difficulty with the HMS, as with the other systems, is the high reliability requirement. A minimum of 45–90 days' mean-time-between-failures may be essential for an efficient HMS. Proof that this can be achieved requires conducting a large-scale system test for a considerable period of time. The propulsion mechanism in a self-propelled collector may need to be refined in order to achieve smooth movement on the sediments. More work needs to be carried out on the positioning of the mining ship, the lift pipe and the collector. Another area requiring investigation is slurry transfer.

Continuous Line Bucket (CLB) Mining System

The CLB is a system that potentially involves relatively less capital costs. It is potentially energy efficient in that the descending line is counter-balanced by the ascending line, whereas in the HMS, large quantities of water are transported. Because the CLB collects whole nodules, it reduces or eliminates separation problems on the surface. Providing the line does not snag, the buckets are frequently available for maintenance. The CLB does not require significant power transmission below the surface either electrically or hydraulically.

However, there are several major difficulties with the system. Among them are:

Potential tangling of the ascending and descending segments of the line.

Potential snagging of the bottom line segment on bottom features. This segment progresses perpendicular to its line of action, and hence, is highly vulnerable to snagging.

Low nodule pick-up by the buckets, which may limit daily production rate.

Lack of control of individual buckets. Buckets may stay on the bottom long after they are filled, and thus, needlessly obstruct uncollected nodules, or buckets may be withdrawn before being filled.

Modular Mining System (MMS)

Under the MMS concept, the production rate could be expanded incrementally by increasing the number of units. Failure of a unit potentially would cause a small percentage change in the production rate, depending on the number of units in use and this effect could be countered by having available an appropriate number of spares. Collector units would be available for inspection on each mining cycle (less than a day in duration) and maintenance could be carried out as required. Moreover, in the event of bad weather the whole mining system could be recovered in one mining cycle. There are economies of mass production arising from the use of identical units. Tailings could be of positive value to mining operations (if environmentally acceptable) and their use as ballast could reduce on-shore tailings disposal costs. However, the high cost of moving tailings from a distant plant site ashore to the mine site could be prohibitive, unless at-sea processing was developed.

The principal technical issue with respect to the MMS is the prospect of loss of the individual units. Another issue is the development of an underwater battery suitable for the system. Systems must be highly reliable and means must be provided to recover failed units. Should the unit loss rate become high, operating costs would rapidly expand. The cost of the units should be such that extensive testing in shallow water and on the deep seabed could be performed before starting production in large numbers.

Chapter 7

Transfer of Nodule Ore at Sea

The decision to transfer nodule ore, personnel, spares and consumables at sea is based on an assessment that it is economically advantageous to do so. Obvious alternatives to transfer at sea are: to merge the transportation and mining functions in one large ship or to process nodule ores at the mine site. However, in these alternatives, the capital investment in mining equipment and in specialized mining system handling equipment, and the capital cost of the configurational impacts on the mining ship design, such as a large moonpool, do not generate revenue during the transportation phase. It is generally accepted that it is economically desirable to continuously operate the mining ship a high percentage of time to maximize the utilization of the capital investment it represents. In this approach, the transport ships could be viewed as essentially conventional bulk carriers which can be leased to reduce capital requirements.

Given a decision to transfer ore and other items at sea, a multi-faceted engineering problem arises. The engineering approach must address the following functions:

Ore de-watering
Buffer storage

Ship positioning
Ship propulsion
Transfer
Queuing

Two key operational parameters are the availability of the mining ship and the availability of the transfer system to the mining ship. The mining ship must have a high availability at the mine site. This is due to the fact that the availability and required production capacity of the mining system are inversely related — the higher the availability of the mining system, the smaller its production capacity requirement. In general, a production capacity for the commercial mining system on the order of

10,000 dmt (dry metric tons) per day is required. To maintain high availability, the mining ship should be able to operate in high sea states — sea state 5 or more — and at an arbitrary heading with respect to the weather. The mining ship needs to operate at arbitrary headings because its trajectory is constrained by the distribution of the nodule deposit and bottom obstacles.

The availability of the ore transfer system must be high; otherwise, both mining ship buffer storage and ore transfer capacity requirements will be high. Thus, the transfer system must also be able to operate in high sea states and with the mining ship at arbitrary headings with respect to the weather.

ORE TRANSFER APPROACHES

Ore transfer approaches are best described in terms of the mode of transfer and the configuration of the mining ship and the transport ship during transfer. Table 6 is a matrix defining the various options with the more feasible options indicated by an "X". The matrix suggests that a slurry transfer mode can be adapted to any mining/transport ship configuration while an alongside configuration allows broadest choice of ore transfer mode. The choice of transfer approach is significantly dependent on the relative advantages/disadvantages of slurry transfer mode versus an alongside configuration. An alongside configuration is a more costly one to implement than others, but the advantages in terms of the flexibility in choice of transfer mode may favour it.

Slurry Transfer Mode

The principal advantage of a slurry transfer is that slurry is the form in which the nodules arrive at the ship. However, this necessitates including de-watering equipment on each transport ship, while if pneumatic or conveyor transfers are used, de-watering equipment is only required on the mining ship. The dynamics and reliability of the slurry hose are major considerations in the design of this transfer mode.

As indicated in Table 6, the slurry transfer mode is adaptable to any of the three configurations; it may, however, be most suitable for the configuration in which the mining ship tows the transport ship. A possible configuration for this is shown in Fig. 17.

A tow length of about 500 m will be required, with a buoyant

Table 6. General Ore Transfer Options

Mining/Transport Ship Configuration in Transfer	Transfer Mode		
	Slurry	*Pneumatic*	*Conveyor*
Mining ship tows transport	x	–	–
Transport tows mining ship	x	–	–
Alongside	x	x	x

hose or hose flotation needed to maintain the hose awash on the surface. The floats and hose may be strapped to a cable which acts as the towing member.

A major concern with the tow configuration is the manoeuvering stability of the mining ship/transport ship system, as the mining operation continues. Thus, thrusters may also be required on the transport ship to minimize the influence of transport ship movement on the positioning of the mining ship.

The slurry transfer mode is not convenient for all other transfer functions such as the transfer of consumables, personnel, and spares. These are more effectively transferred in an alongside configuration. Hence, both configurations may be required when the slurry transfer mode is used and nodules may have to be kept in buffer storage during the transfer of the items other than ore.

Alongside Configuration

The alongside configuration allows a wider choice of transfer mode as indicated in Table 6. The major issue is the positioning of

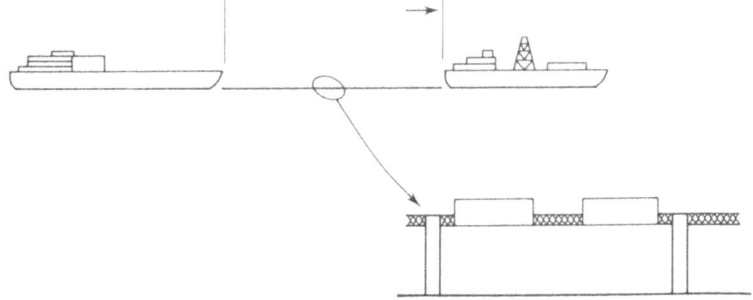

Fig. 17. Mining ship tows the transport ship.

the two ships. Several methods can be envisaged for maintaining the relative positions of the two ships and for manoeuvering them in tandem. These are categorized as follows:

Position-keeping in close proximity.
Rigid structural connection with pivoted joints.
Lashings and fenders.

The major concern in the position-keeping method is the possible collision of the two ships. Thus, the ships may have to stand-off at least 300 m. The feasible ore transfer modes are pneumatic or slurry transfer via a hose. The pneumatic transfer allows the de-watering equipment to be located on the mining ship, and hence, if feasible, it is the preferred transfer mode when position keeping is used. This method requires each transport ship to be equipped with thrusters and on-board computer control of the transport ship's position relative to the mining ship.

A second possible method in alongside transfer is to use a rigid structural frame which precludes some components of relative motion (relative sway, surge, and yaw) while permitting others (relative heave, roll, and pitch. (Schematically, such a system is shown in Fig. 18. In order to limit transverse bending loads in the members, universal joints are used as connections with the ships and the connector has axial force components. The structure is connected so that the induced forces produce minimum roll effects on the two ships. Since large compressive forces are produced in the components of a rigid frame, structural requirements may be minimized by shortening the components and reducing the separation distance of the two ships. This, on the other hand, increases the risks of collision. A refinement of this method is to use telescoping components with adjustable stiffness and damping (an air spring, for example) so that the connection can be stiffened as it is shortened.

The rigid structural connection or a method using lashings and fenders maintains the two vessels in close proximity, and thus, provides the maximum flexibility for the choice of transfer mode. However, a number of technical difficulties are involved in the design of the interconnection.

The selection and development of a transfer-at-sea approach involves many interdependent decisions. Thus, it is best addressed by a systems engineering approach which includes development of designs, technical requirements, and cost information on various approaches.

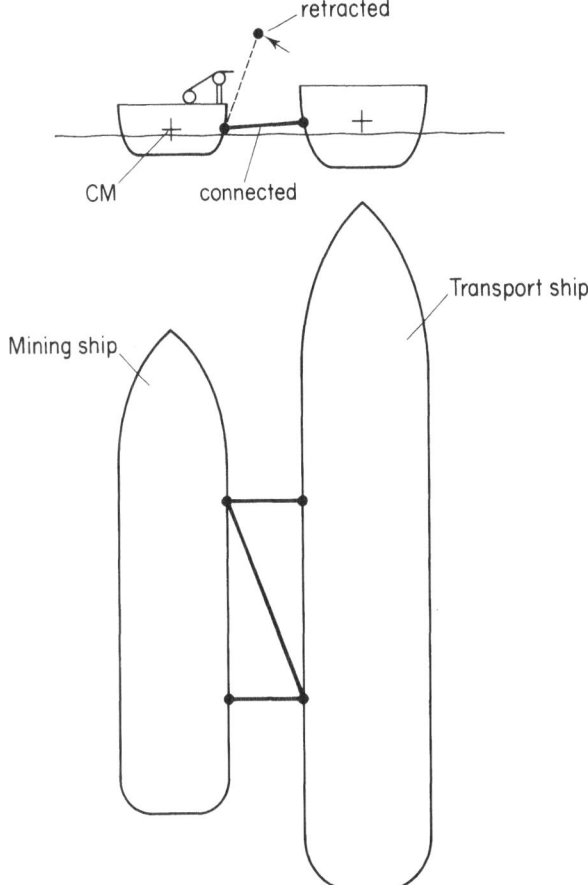

Fig. 18. Schematic of a rigid structural connection between mining ship and transport ship.

The transfer-at-sea problem in nodule mining is unique in scope because of several factors:

The rate at which material is to be transferred is very high— in the range of 10,000 dmt/day; the transfer system must also provide for transfer of personnel, consumables and spares.

Transfer must occur at low speed, in all weather headings, and during relatively high sea states.

Motion of the transport ship should not hinder the mining operation.

Figure 19 presents a breakdown of the components of ore transfer technology.

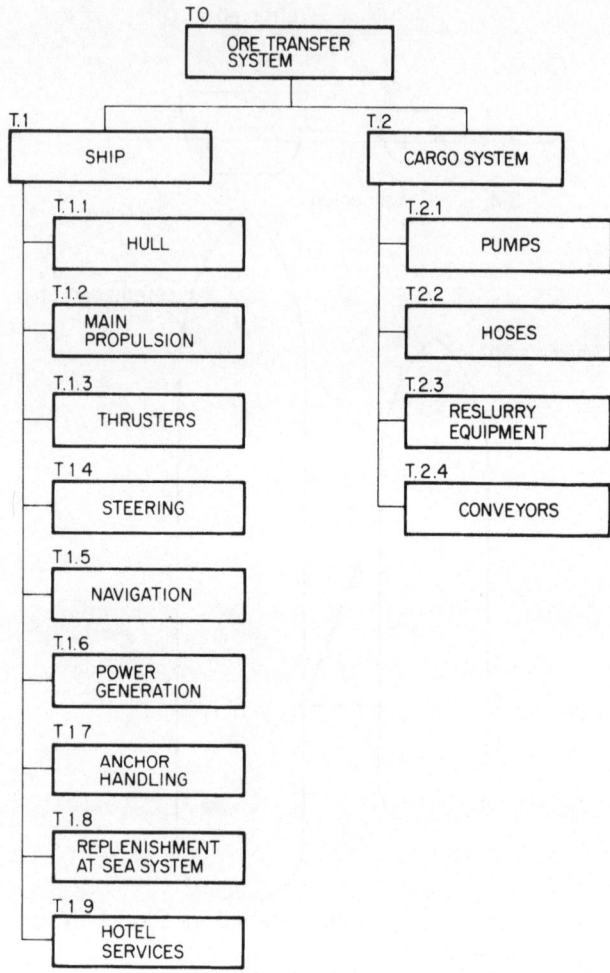

Fig. 19. Ore transfer technology component breakdown. (Reprinted with permission from Science Applications, Inc. and United States Bureau of Mines.)

As a final comment, it should be pointed out that experience with high-volume transfer of solid materials in the open ocean has been extremely limited. During naval exercises liquid fuel has been pumped from ship to ship and some solid supplies transferred by line. It is known that position keeping can be difficult even in moderate sea states. Ore transfer in high volume poses a major problem and requires experimentation that has yet to be attempted by the developers.

Chapter 8

Conclusions

After reviewing the technology of seabed mining that has been developed in the last decade or so, it can be concluded that the pioneer groups made a significant contribution. At the end of a relatively short period of time, and with somewhat limited financial outlays, they have made remarkable advances in seabed exploration, development of submersible mining equipment and metallurgical research. Their work has removed most of the scepticism that existed in the 1960's and early 1970's. Although the task is far from being completed, one can envision the possibility of commercial operations using first-generation equipment and methods. The current pause in test work is not because of technical factors, but rather because of existing low metal prices and a depressed world economy. If the next swing in the business cycle is upward, bringing with it an increase in the worldwide demand for metals, a resumption of seabed mining activity can be expected.

It is expected that future exploration, especially detailed mapping, will seek to employ more extensively improved TV equipment, narrow-beam up- and down-looking sonars, side-scan sonars and other refined instruments so as to give a more accurate picture of the microtopographic features identifying minor obstacles and, hopefully, providing a view of the nodule fields over a greater extent of the seafloor.

At present it is impossible to determine whether or not, or when, advances in technology will reach a state where equipment, both sub-surface and on the mining ship, will achieve higher reliability as well as efficiency. More skilful exploration and mining methods will probably come about with greater experience in the deployment and recovery of submerged apparatus. Technology should also improve as the developers gain some experience in the collection and transfer of ores to a carrier. More reliable collectors and improved lift systems should evolve with experience. Similarly, the pumping and dewatering systems used in ore transfer will be further investigated with a view to improving them. Better weather reports and experience in operating at different sea states should lead to practised response by the miners and greater ore productivity.

Considering the technical progress that has taken place in other industries over the years, there seems good reason to be encouraged that this nascent industry can benefit dramatically with continued experimentation.

Continued experimentation will next involve at-sea mining tests that will be conducted over a much longer period than heretofore. The mining tests to date have been of short duration, according to public reports. These reports suggest that the tests have been somewhat discontinuous and conducted over a period of only a few days. The next series of tests, to be meaningful, will have to be conducted over a period of many days or a few weeks. The purpose of these longer-term tests would be to demonstrate the reliability of the mining system in a time frame that represents a significant segment of a normal commercial operation. Major financing to undertake a commercial venture will probably not be forthcoming until such a demonstration has occurred.

Looking ahead, the successful completion of the exploration sequence accompanied by progress in mining and processing technology will mark the progression of manganese nodules from a mineral resource to a mineral reserve. The nascent industry may then evolve into a multi-billion dollar industry.

Abstracts of Patents on Seabed Exploration and Mining Technology[13]

US 3,588,174 — Apparatus and method of using it for mining manganese nodules or other loose mineral lumps and pieces from the ocean floor at considerable depth. The foldable arms of the apparatus are opened and pulled along the ocean floor. The mineral pieces are gathered hydraulically and directed to an intake, and moved from the intake to a surface vessel via an upwardly flowing stream of water in a conduit. A.M. Rossfelder and B.J. Thorn, assigned to Tetra Tech, Inc. June 28, 1971. 10 pp. (299-8).

British 1,239,178 — In the mining of manganese nodules or other particulates from the ocean floor at depths of up to 4 miles, a long endless chain, rope, or wire loop having buckets attached thereto at intervals is dropped from the stern of the vessel and raised at the bow of the vessel, so as to scrape the ocean floor before being raised. The vessel is moved sideways so that the distance between the descending part and the ascending part of the loop is kept at a maximum. Loaded buckets are dumped into the vessel before again descending. Y. Masuda and H. Fukada. July 14, 1971. 9 pp. (E 02 F 3/08).

Canadian 878,952 — In the recovery of manganese nodules or other lump or particulate mineral from the seabed at considerable depth, conduit sections, connected to an underwater collecting apparatus, are laid out behind a moving vessel so as to assume a desired angle at the same angle as that of a derrick on the vessel. When the string of conduit sections reaches the mineral formation on the bottom, the vessel is slowed down to a predetermined speed for recovery operations. M.W. Smith and C.S. Kluth, assigned to Westinghouse Electric Corp. Aug. 24, 1971. 17 pp. (262-27). Same: **US 3,543,527,** dated Dec. 1, 1970.

British 1,243,615 — In the recovery of manganese nodules or other seabed minerals from depths of up to 12,000 feet, the vessel is positioned over the area to be mined, and a series of conduit sections having a collecting head on the lowermost conduit is paid out vertically until the collecting head engages the mineral bed. The lowermost sections are made pivotable so that the collecting head may operate to gather mineral pieces. The gathered mineral is pumped through the conduit sections and into the vessel for beneficiation and/or transportation to shore facilities. J.E. Flipsie, *et al.*, assigned to Newport News Shipbuilding and Dry Dock Co. Aug. 25, 1971. 7 pp. (E 02 F 3/88; E 02 F 5/28).

Canadian 888,589 — Collecting device and method of using it for selectively mining manganese nodules, phosphate nodules, or like aggregates of a preselected size range only from the ocean floor at considerable depths, even over undulating or irregular surfaces and without need for electrical or movable mechanical connections with a surface vessel. A scoop rotates around an axis to pick up the solid material and deposit it in a hopper. The scoop has screens for eliminating oversize lumps and undersize particles. M.W. Smith, assigned to Westinghouse Electric Corp. Dec. 21, 1971. 27 pp. (262-31).

British 1,262,522 — In the sampling of ocean floor deposits of manganese nodules or like lump or particulate material, a frame having a pair of open jaws operable to close upon contact is dropped to the floor, whereupon the open jaws are closed to retain a sample of nodules, a ballast means is released, and the sample is buoyed to the surface and picked up. This is a free-fall apparatus; a retrieving line to the launching vessel is not required. Assigned to Bear Creek Mining Co. Feb. 2, 1972. 14 pp. (E 02 F 3/44). Same: **US 3,572,129,** dated Mar. 23, 1971.

British 1,262,660 — In the recovery of manganese nodules or other particulate lump, or vein, material from the floor of an ocean or lake, a closed chamber is positioned over a portion of the mineral bed or vein, a solvent or other reagent supplied to the chamber to react with the ore, and the material in the chamber subjected to *in-situ* electrolysis whereby the metal values are recovered. F.W. Wanzenberg and Fritz W. Wanzenberg. Feb. 2, 1972. 17 pp. (E 21 B 43/91).

Canadian 823,780 — In the collecting of samples of manganese nodules or other loose particulate minerals from the floor of the ocean, jaws held in an open condition by a release means and mounted on a framework are lowered to the seabed, the release means actuated by contact with the seabed to enable the jaws to close and thereby take the sample, and the sample thus taken is buoyed to the surface. T.N. Walthier, C.E. Schatz and A.M. Rossfelder, assigned to Bear Creek Mining Co. Feb. 22, 1972. 33 pp (73-102). Same: **US 3,572,129**, dated Mar. 23, 1971.

US 3,672,079 — In the mining of manganese nodules from the ocean floor at considerable depth, an endless loop rope-and-bucket system is operated from a vessel moving slowly on the surface of the ocean, the loop being laid out at the bow of the vessel and lifted at the stern for dumping of the buckets into a hold. The buckets are in fact open wire nets mounted on frames, the nets being of mesh small enough to retain nodules while passing the usual bottom mud. Y. Masuda and T. Murakami. June 27, 1972. 6 pp. (37-69).

US 3,672,725 — In the mining of manganese nodules from the ocean floor at considerable depth, the mineral-bearing area is traversed by a mining vehicle operated from a mother vessel. Collected nodules are crushed and classified by the vehicle to form a mixture having a predetermined solid/fluid ratio range, and the mixture pumped through a riser conduit to the surface vessel. E.P. Johnson, assigned to Earl and Wright. June 27, 1972. 16 pp. (299-8).

US 3,675,348 — In the recovery of manganese nodules or other loosely embedded lump minerals from the ocean floor at considerable depth, an endless chain mounted on a long, foldably extensible, double-tiered track and operated by an undersea vehicle carries a series of buckets which scrape loose and retain the lumps, bringing them back to the vehicle for pumping to the surface. To reduce crabbing, the buckets are skewed forward of the vehicle at a predetermined optimum angle relative to the axis of the track. E.B. Dane, Jr. July 11, 1972. 14 pp. (37-69).

US 3,697,134 — Mechanical collector and method of operating it for gathering manganese nodules or the like from the ocean floor. The apparatus is lowered to the ocean floor and towed along by a floating vessel as it sweeps nodules up a grating and into a basket. The gathered nodules may be either pumped to the vessel via a connecting pipe or hoisted using a cable connected to the apparatus. R.H. Murray, assigned to Bethlehem Steel Corp. Oct. 10, 1972. 8 pp. (200-8).

Canadian 920,619 — Method and related apparatus for recovering values from deposits of manganese nodules or other particulate material containing metal values and located on the ocean floor at considerable depths. The values are extracted from the material at the underwater site of the deposit, using solution chemistry, solvent extraction, and electrodeposition. The resulting enriched granules or slimes are then removed to the surface for further processing. Handling of large amounts of gangue material is avoided. F.W. Wanzenberg and Fritz W. Wanzenberg. Feb. 6, 1973. 39 pp. (262-9). Same: **US 3,748,248,** dated July 24, 1973.

Canadian 928,337-8 — These two patents cover a method and apparatus for mining ocean bed deposits of manganese nodules or the like from depths of more than 400 feet. Gathered nodules are pumped via joined conduit sections to the surface vessel, where they are either processed or accumulated pending transportation to a processing facility. N.D. Birrell, *et al,* assigned to Newport News Shipbuilding and Dry Dock Co. June 12, 1973. 18 pp. and 16 pp., respectively. (262-27).

US 3,753,303 — Mining of manganese nodules or other particulate solids from the floor of the ocean. Ore collected by a suitable pick-up apparatus is fed into a riser conduit provided with from two to four containers, which containers are utilized cyclically for producing a pressure differential for raising the ore–water mixture in such a manner that it flows continuously from the upper portion of the riser conduit. K. Holzenberger and O. Schiele, assigned to Klein, Schanzlin and Becker A.G. Aug. 21, 1973. 14 pp. (37-58).

US 3,765,727 — Process and apparatus for moving to a surface vessel manganese or phosphate nodules mined from the ocean floor at considerable depth. The air is withdrawn from an air lift hydraulic pipe at one or more points substantially above the air injection station, whereby velocity and turbulence in the upper section of the pipe are controlled and the exit velocity of the mined material from the pipe is restricted. Nodule breakage and pipe wear are lessened. J.C. Santangelo, M.A. Dubs and C.E. Schatz, assigned to Kennecott Copper Corp. Oct. 16, 1973. 8 pp. (302-14).

US 3,776,593 — Apparatus and related method for collecting manganese nodules from the floor of the ocean. A toothed rotating drum having inside it a plurality of stationary arcuate magnets arranged side-by-side with north-south poles alternating with each other in the axial direction of the drum is pulled over the ocean floor upon which nodules are lying, so that the nodules are magnetized as the magnets pass thereover. The magnetized nodules are picked up by the teeth and lifted to and deposited in a suction intake feeding to a conventional conveyor tube. W.H. Kuhlmann-Schaefer and M.E. Dinter, assigned to Preussag A.G. Dec. 4, 1973. 10 pp. (299-8). Same: **British 1,387,694,** dated Mar. 19, 1975.

US 3,777,377 — Recovery of manganese nodules or like mineral materials from the ocean floor at considerable depths. Two buckets are raised and lowered alternately by a winch from a meandering vessel, with one bucket gathering nodules while the other is being emptied into the hold of the vessel. M. Toritani. Dec. 11, 1973. 6 pp. (37-195).

US 3,802,740 — Recovery of manganese nodules or other mineral lumps from the ocean floor at depths up to 3 miles. An apparatus having means for gathering and at least partly de-silting ore pieces is pulled along the ocean floor. Collected material is lifted a short distance into a partially enclosed compartment which permits silt to be washed out. The ore pieces are raised to a surface vessel via a conventional conduit device. A.F. Sullivan, assigned to International Nickel Co., Inc. Apr. 9, 1974. 9 pp. (299-8). Same: **British 1,405,997** dated Sept. 10, 1975 and **Canadian 994,819,** dated Aug. 10, 1976.

US 3,811,730 —Mining of manganese nodules or other loose ore sediment from the ocean floor at depths of 100 fathoms or more. Ore collected from the deposit by a scraper bucket system is directed into a sub-sea apparatus which cleans the ore of mud, crushes the cleaned ore, and introduces the crushed ore into a hoist pipe circuit wherein it is lifted as slurry and deposited in bins on a hovering "Sessile" ship. E.B. Dane, Jr. May 21, 1974. 32 pp. (299-8).

US 3,812,922 — Mining of manganese nodules or other bottom sediment mineral material from the ocean floor at considerable depth. A mining vehicle provided with variable buoyancy tanks and mining mechanisms is propelled along the floor of the ocean by means of high velocity jets and/or turbine wheels. Mineral pieces are pumped to a payload receiving chamber, and the bulk of the water and sediment removed. The vehicle operates through its mining cycle completely independently of the mother ship. B.G. Stechler. May 28, 1974. 19 pp. (175-6).

US 3,829,160 — Bucket dredging of manganese nodules from the seabed. Nodules are scraped up into a current of washing water produced by the forward movement of the dredging bucket whereby bottom clay is washed from the nodules, the clay-free nodules moved into a central hopper, and nodules from the hopper pumped to the surface vessel. E. Condolios, assigned to Ste. Gen. de Constructions Electriques et Mecaniques (Alsthom). Aug. 18, 1974. 5 pp. (299-8).

US 3,842,522 — Hydraulic system and method for raising from the sea bottom a plurality of containers bearing manganese nodules or other mineral pieces or particulates. Raising is effected in a riser conduit using a flushing pumping means and a main suction pump means which do not have contact with the ore. Clear flushing water is exhausted from one container, whereby a pressure differential is produced in a lower portion of the riser conduit. K. Holzenberger and O. Schiele, assigned to Klein, Schanzlin and Becker A.G. Oct. 22, 1974. 14 pp. (37-195). Same: **British 1,372,866,** dated Nov. 6, 1974.

British 1,373,890 — Method and apparatus for extracting manganese nodules or the like from the ocean bed at considerable depth, using two ships travelling parallel courses. An endless line of buckets is fed down to the deposit from the first ship and passed along the ocean bed, the filled buckets raised to the second ship and dumped, and the emptied buckets returned to the first ship along a path which is at least partly submerged. Navigation problems are minimal, and operation parameters – such as changes in depth of the deposit – can be adjusted to readily. Assigned to Centre National pour l'Exploitation des Oceans and Le Nickel. Nov. 13, 1974. 4 pp. (E 02 F 3/08). Same: **US 3,889,403,** dated June 17, 1975 and **Canadian 970,399,** dated July 1, 1975.

Canadian 959,506 — System for recovering manganese nodules from the ocean bed at considerable depth. A magnetically non-conductive drum provided with radially projecting rims is rolled along the ocean bed to stir up and magnetize the nodules, and the magnetized nodules are retained by a magnet or magnets and deposited on a conveyor. W. Kuhlmann-Schafer and M. Dinter, assigned to Preussag A.G. Dec. 17, 1974. 19 pp. (262-28).

British 1,382,388 — Method and apparatus for obtaining samples of manganese nodules or other mineral particulates from the ocean floor. A free-grab device mounted on a rigid frame is ballasted and locked into open position, the ballasted device lowered to the ocean floor, and the jaws swung outwardly to trap the material being sampled by trapping it with the jaws abutting against the frame. The ballast is unloaded, and the device hoisted to the ship. Assigned to Centre National pour l'Exploitation des Oceans, Le Nickel, and Tetra Tech Inc. Jan. 29, 1975. 5 pp. (E 02 F 3/44).

US 3,908,290 — Recovery of manganese nodules from deep sea beds. A first vessel for receiving nodules has a pivotally connected nodule removing tube extending approximately vertically to the floor of the ocean, where a nodule gathering mechanism is towed and operated from a second vessel. The distance between the vessels may be increased to thereby effect at least partial raising of the tube during movement of the second vessel. E. Condolios, assigned to Ste. Gen. de Constructions Electriques et Mecaniques (Alsthom). Sept. 30, 1975. 9 pp. (37-58).

US 3,908,291 — Mining of manganese nodules from deep sea beds, using a single ship of normal size and buckets affixed to an endless rope. To prevent tangling of the descending and ascending portions of rope, each bucket is provided with a planar plate having a hydrofoil cross-sectional configuration. As the ship moves through the water, the bucket plates tend to prevent dragging of the descending rope portion, thereby maintaining adequate distance from the ascending portion to avoid tangling therewith. Y. Masuda. Sept. 30, 1975. 17 pp. (37-69). Same: **Canadian 997,381,** dated Sept. 21, 1976.

US 3,942,003 — Method and apparatus for the *in-situ* analysis of manganese nodules on the ocean floor, prior to recovery thereof. Nodules are raised into an apparatus and de-silted, the de-silted material exposed to radioactive rays for inducing a secondary radiation emanating from the material, and the secondary radiation analyzed to determine the quantity of the individual elements contained in the nodules. The bulk density of the material also may be ascertained in the apparatus. W. Apenberg, G. Bohme, H.-U. Fanger, B. Glaser, K. Hain, J. Hubener, W. Stegmaier, V. Prech and J. Vagner, assiged to Gesellschaft fur Kernforschung m.b.H. Mar. 2, 1976. 13 pp. (250-255). Priority: West Germany, 2-28-73 (2,309,974).

US 3,943,644 — Apparatus and method for dredging manganese nodules from the ocean floor. A flexible combined guide train and conveying train assembly is suspended between an on-ship drive unit and a receiving unit which is towed along the ocean floor to scoop up the nodules. The system is also useful for recovering mineral silt. A. Walz. Mar. 16, 1976. 24 pp. (37-69). Priority: West Germany, 6-25-73 (2,332,198).

US 3,947,980 — Recovery of manganese nodules, phosphate nodules, mineral "ooze", or sedimentary tin, gold, platinum or other values from the ocean floor. Values are collected by a sled towed along the seabed by a single towing strip that includes a bucket elevator mechanism. The buckets load from a concentrator collector on the sled and empty onto a surface vessel. This apparatus easily adapts to elevational changes between the surface vessel and the towed sled caused by ocean swells or changes in the level of the ocean bottom. J.B. Andrews and M.E. Morganstein, assigned to Hawaii Marine Research, Inc. Apr. 6, 1976. 10 pp. (37-69).

US 3,968,579 — Apparatus and related method for gathering manganese nodules or like sediments from the bottom of the ocean. A continuous hollow loop element is operated from a surface vessel to the ocean bottom and return. The loop has a ballasted fluid in a lower portion, while the descending portion is buoyant to lessen the likelihood of entanglement. A weight sled is used to help gather nodules and prevent wear of the loop. Gathered nodules are raised via buckets. A.M. Rossfelder. July 13, 1976. 10 pp. (37-69).

US 3,971, 593 — Recovery of manganese nodules from the seabed. Nodules are dredged and raised to a subsurface control and storage station, the nodules de-aerated, the separated air re-injected into the dredging tube, and the de-aerated nodules pneumatically lifted to the surface vessel. R. Porte and M. Rommens, assigned to Commissariat a l'Energie Atomique. July 27, 1976. 6 pp. (299-8). Priority: France, 7-18-73 (73/26289). Same: **Canadian 1,012,566,** dated June 21, 1977 and **British 1,464,184,** dated Feb. 9, 1977.

US 3,972,566 — Method and apparatus for concentrating deep sea manganese nodules or other mineral solids from the ocean floor. Solids are gathered into a hollow concentrator drum as it is moved along the seabed, the gathered solids passed through the drum and sized, and the pieces of selected size either rolled into a lifting conduit or windrowed on the seabed for subsequent collection by another apparatus. F.H. Brockett, III, and A.F. Sullivan, assigned to International Nickel Co., Inc. Aug. 3, 1976. 8 pp. (299-8). Same: **Canadian 1,018,557,** dated Oct. 4, 1977 and **British 1,518,426,** dated July 19, 1978.

US 3,973,575 — Apparatus for collecting and concentrating deep sea manganese nodules or the like, and for moving the concentrated nodules into a conduit for lifting to a surface vessel. The vehicle has two horizontally fenestrated sweeps provided with horizontal bars which concentrate desired sizes of solids into a windrow, a transition tailpipe for picking up the windrowed solids, and an exit through which solids and water are directed into the suction conveyance duct. This vehicle is towed along the seafloor. A.F. Sullivan and F.H. Brockett, III, assigned to International Nickel Co., Inc. Aug. 10, 1976. 8 pp. (37-58). Same: **Canadian 1,018,192,** dated Sept. 27, 1977.

US 3,975,054— Apparatus and method for recovering manganese nodules or the like from the ocean bottom. The towed vehicle has a motor-powered water jet, an enclosed ramp, with an entrance behind the water jet, a hopper and trough to receive dislodged solids, and separating screens and ducting to direct sized solids into a conduit for delivery to a surface ship. F.H. Brockett, III, J.E. Philip and A.F. Sullivan, assigned to International Nickel Co., Inc. Aug. 17, 1976. 12 pp. (299-8). Same: **Canadian 1,018,558,** dated Oct. 4, 1977.

US 3,975,841 — Design for and method of operating a mining ship for recovering manganese nodules or other mineral solids from the seabed. The ship carries mineral recovery apparatus and has an adjoining unitarily connected pier to which may be moored at least one transport ship into which the recovered mineral material is loaded. The transport ship or ships provide part of the propulsion force for the mining ship. H.W.L. Steenken and D.G.H. Luth, assigned to Howaldtswerke-Deutsche Werft A.G. Hamburg und Kiel. Aug. 24, 1976. 6 pp. (37-54). Priority: West Germany, 4-20-74 (2,419,056). Same: **Canadian 1,009,262,** dated Apr. 26, 1977 and **British 1,460,877,** dated Jan. 6, 1977.

British 1,447,310 — Method and apparatus for extracting manganese nodules, ore sludges, or the like from the ocean floor. Buckets are fixed at predetermined distances on an endless rotatable conveyor suspension gear. The gear is driven from a driving unit aboard a surface vessel to a receiving unit positioned on a gathering unit which traverses the seabed collecting the material being extracted. The gear has guides for controlling the forward and return strand runs of the conveyor suspension device from the receiving unit. A. Walz. Aug. 25, 1976. 23 pp. (E 02 F 3/08). Priority: West Germany, 6-25-73 (2,332,198).

US 3,988,843 — Apparatus and related method for injecting manganese nodules or other seabed mineral particulates into a flowing stream of water in a hydraulic suction conduit through which the particulates are lifted to a surface vessel. Nodules collected by a gathering vehicle enter through a forward-facing opening, and water through a second opening. The nodules and water are mixed, and the suspension injected into the conduit via an exit port. F.H. Brockett, III, assigned to International Nickel Co., Inc. Nov. 2, 1976. 7 pp. (37-57).

US 3,999,313 — Apparatus and method of operating it to recover manganese nodules or other particulate or lump material from the seabed. A sled supported by spaced-apart runners is towed over the seabed, and material is collected in an open container. Cable mounted buckets are moved continuously through the container to scoop up gathered material and transport it to the surface vessel. J.E. Andrews, assigned to Hawaii Marine Research, Inc. Dec. 28, 1976. 11 pp. (37-69).

British 1,464,332 — A surface installation for handling manganese nodules recovered from the ocean floor by the method described in US Patent No. 3,975,841, dated Aug. 24, 1976, or similar method includes a self-propelled buoyant vessel having at least one lighter-receiving chamber provided with an overhead bunker fixed along its longitudinal axis. Recovered nodules are delivered via the bunker into the lighter in the chamber, either direct or after preliminary beneficiation on the vessel. Assigned to Howaldtswerke-Deutsche Werft A.G. Hamburg und Kiel. Feb. 9, 1977. 5 pp. (B63B 35/00; B63B 35/28). Priority: West Germany, 4-20-74 (2,419,057).

British 1,464,549 — Apparatus and method for collecting particulate solids, such as manganese nodules, from the seabed or other relatively deep underwater locations. A large closed hollow structure in the shape of an enclosed diving bell is lowered to the location of the solids, and the solids are drawn by means of suction into the structure through peripheral inlets. As the structure fills with solids, air and water are forced out of the structure. The loaded structure is caused to float, and the solids are recovered from the floating structure. R.A. Nixon, assigned to Secretary of State for Industry. Feb. 16, 1977. 8 pp. (E 02 F 3/88).

US 4,010,560 — Apparatus and method for mining manganese nodules from the ocean floor. A plurality of nodule harvesting machines is operated along the ocean floor from a surface ship. Each machine includes separable crates which, when filled with nodules, are lifted by cable to the surface vessel and emptied, and the empty crates returned to the machine. Placement and guidance of the harvesting machines are controlled by sonar devices and television cameras. R.E. Diggs. Mar. 8, 1977. 12 pp. (37-54). Same: **British 1,533,822,** dated Nov. 29, 1978 and **Canadian 1,047,546,** dated Jan. 30, 1979.

US 4,019,380 — Method and apparatus for sampling deposits of manganese nodules on the ocean floor. A free-fall sampler comprising a pair of jaw members and ballast supporting containers which are operated independently of one another is dropped to the deposit site, the jaws caused to close, a portion of the ballast released, and buoyant means on the frame activated to cause the sampler to rise to the surface with the enclosed nodules. S.O. Raymond, assigned to Benthos, Inc. Apr. 26, 1977. 14 pp. (73-170A).

British 1,472,163 — Samples of manganese nodules or other lump or particulate materials are recovered from the seabed by lowering onto the bed a free-fall grab device having a buoyancy member positioned above pivotable grab halves, closing the grab halves around the sample, and releasing the buoyancy member to raise the apparatus. Assigned to Preussag A.G. May 4, 1977. 7 pp. (E 02 F 3/44). Priority: West Germany, 9-16-74 (2,444,167).

US 4,030,216 — Hydraulic system for lifting manganese nodules or other particulates from the ocean floor. A partially gas-filled submerged chamber is suspended from a stable surface platform and connected to the open environment by means of pipes connected to a buoy. The chamber is located at such a depth that the pressure difference between the inside of the chamber and the surrounding water is greater than the pressure drop caused by the transport of ore and water through the pipe from its lower end to the chamber. The mixture of ore and water is subsequently transported in buoyant containers, guided by a paravane. J.-O. Willums, assigned to Nor-Am Resources Technology, Inc. June 21, 1977. 7 pp. (37-58).

US 4,035,022 — In the recovery of manganese nodules or the like from the seabed, a self-propelled pickup device having a cutting tool and a conveyor apparatus is moved over the seabed. Particulate material is dug from the seabed, and forced onto a conveyor and from the conveyor trough into an elevating chain-type conveyor. The conveyor delivers to an inclined classifying screen, and desired material is retained on the screen and passed into a hopper leading to an elevator. U. Hahlbrock, S. Steinkuhler, H.-W. Vagt, J.-G. Voss and U. Wantig, assigned to Orenstein and Koppel A.G. July 12, 1977. 10 pp. (299-8). Priority: West Germany, 2-5-75 (2,504,694). Same: **Canadian 1,039,760,** dated Oct. 3, 1978.

US 4,037,874 — Manganese nodules or like particulate is recovered from a smooth or uneven seabed by moving along the seabed a rotatable drum having a plurality of rows of digging forks and a plurality of adjacent rows of openings for receiving material moved by the forks. The size of the openings determines the maximum size of pieces which will be recovered. Recovered material is moved longitudinally in the drum via a screw trough to a pick-up region, from which the material is transported hydraulically to the surface. J.-O. Willums, assigned to Nor-Am Resources Technology, Inc. July 26, 1977. 4 pp. (299-8).

US 4,040,667 — Manganese nodules are recovered from the seabed by means of a dredge mechanism associated with a transfer station which is suspended from a mother ship at a fixed depth. The dredging tool assembly is suspended from the transfer station via cables whose lengths are varied in response to depth signals from sensors, to thereby maintain the dredge in effective contact with the seabed. Gathered material is processed through the transfer station, and the de-silted, sized nodules are lifted to the mother ship through an hydraulic conduit. H. Tax and H. Kurz, assigned to Hans Tax. Aug. 9, 1977. 11 pp. (299-8). Priority: West Germany, 8-16-74 (2,439,485). Same: **Canadian 1,030,163,** dated Apr. 25, 1978.

US 4,042,279 — Recovery of manganese nodules or the like from the seabed. A gathering apparatus including a frame having slide bodies on its bottom surface is driven along the seabed. Material is loosened by water jets, the loosened material gathered in a collection chamber, and material from the collection chamber transported to a mining ship via suction pipe. Y. Asakawa, assigned to Sumitomo Metal Mining Co., Ltd. Aug. 16, 1977. 9 pp.

(299-8). Priority: Japan, 10-2-75 (50/119224). Same: **Canadian 1,040,224,** dated Oct. 10, 1978 and **British 1,545,867,** dated May 16, 1979.

Canadian 1,017,759 — In the recovery of manganese nodules or other particulate material from the seabed, gathered material is passed to a transition chamber mounted on the aft end of the gathering apparatus. In the chamber, the material is de-silted and sized. Particulates of selected size are introduced into a hydraulic suction conduit and raised to a surface vessel. F.H. Brockett, III, assigned to Inco Ltd. Sept. 20, 1977. 16 pp. (262-27). Priority: US 12-11-74 (531,753).

US 4,052,800 — Manganese nodules or other loose particulate solids are gathered from the seabed by dragging across the seabed an apparatus having a relatively large inlet and small outlet so as to form windrows of solids, and the windrows are drawn into a separate suction conveyor for lifting to a surface vessel. A. Fuhrboter and M. Mittelstadt, assigned to Salzgitter A.G. Oct. 11, 1977. 7 pp. (37-58). Priority: West Germany, 8-1-74 (2,437,071). Same: **British 1,512,808,** dated June 1, 1978.

US 4,053,181 — In the recovery of manganese nodules, an open bottom hollow box tank suspended from a ship is trawled along the seabed, a suction force maintained in the box by a pump on the ship connected to the box tank via a suction conduit, and the mixture of sea water and entrained solids is conveyed to the ship via the conduit. N. Saito. Oct. 11, 1977. 8 pp. (299-9). Priority: Japan, 1-20-76 (51/5583).

US 4,055,006 — Manganese nodules or the like are collected from the seabed and placed in baggy nets fitted along a rope or chain hanging from a surface vessel, and the nets lifted to the vessel. The collecting plow device is towed along the seabed ahead of the lifting mechanism, and the empty nets are lowered to a trailing sled and then passed to the rear end of the collecting device for filling. J. Shibata, assigned to Mitsubishi K.K.K. Oct. 25, 1977. 6 pp. (37-69). Priority: Japan, 9-21-73 (48/106590) and 10-3-73 (48/111111).

US 4,070,061 — Manganese nodules are recovered by traversing the ocean floor with a dragged or self-propelled unit having shoes for dislodging nodules into a duct having nozzles for projecting a liquid. The duct increases in cross-section, thereby slowing the velocity of the suspension of nodules and silt, causing substantial separation thereof. Nodules are removed from the duct upstream of the final outlet through which silt and water are ejected. V. Obolensky, assigned to Union Miniere. Jan. 24, 1978. 8 pp. (299-8).

US 4,085,973 — Manganese nodules or other mineral particulates are recovered from the floor of the ocean by means of an apparatus having driving means for at least partially penetrating the ocean floor, means for buoying the penetrating mechanism out of the ocean floor and returning it to the ocean surface, and expandable means for trapping the loosened mineral particulates of predetermined size and raising them to the ocean surface. Apr. 25, 1978. 7 pp. (299-8).

US 4,141,159 — Recovery of manganese nodules from the ocean floor. A harvesting apparatus is suspended from a ship by a string of dual concentric pipes containing pressurized water for lifting the nodules to the surface. Nodules are dislodged from the seabed by jets of water, the dislodged material pushed up a channel to a roller crusher on the harvesting apparatus, and the crushed nodules stored in a bin until fed by gravity into the lifting system. W.V. Morris and G.W. Sheary, III, assigned to Summa Corp. Feb. 27, 1979. 13 pp. (37-58).

US 4,147,390 — An apparatus for dredging manganese nodules from the seabed includes a convergent wall device, a gridlike lifting device, and a nodule collection chamber with an inclined surface. A flow of water and nodules is driven up an inclined surface and cleaned, a flow of water and dispersed silt projected upward and rearward, and the nodules collected in a chamber having an entrance turned away from the direction of silt flow. J. Deliege, M. Giot, V. Obolensky, L. Deconinck, and M. Lejeune, assigned to Union Miniére. Apr. 3, 1979. 29 pp. (299-8). Priority: Luxembourg, 8-6-75 (73,155). Same: **British 1,551,926**, dated Sept. 5, 1979.

US 4,147,454 — Method and apparatus for the *in situ* construction of large diameter pipes for use in lifting manganese nodules or the like from the ocean floor. One or more rolls of previously prepared sheet materials are wound around a core in a crosswise fashion and bonded and treated by chemical and physical processes on a special floating platform in the immediate neighbourhood of the mining operation. Formed pipe may be stored or moved directly into the ocean. J.-O. Willums, assigned to Nor-Am Resources Technology, Inc. Apr. 3, 1979. 5 pp. (405-156).

US 4,149,739 — In the recovery of manganese nodules from the ocean floor via a string of pipe connecting the surface vessel to the nodule harvesting apparatus, the pipe is made of concentric inner and outer pipes. The outer pipes are screwed together in fixed relationship, whereas the inner pipes telescope and are suspended at their upper ends by a rigid supporting ring supported by an internal shoulder of the outer pipe. Thread corrosion is prevented. W.V. Morris, assigned to Summa Corp. Apr. 17, 1979. 10 pp. (285-133R).

British 1,553,477 — An apparatus for mining manganese nodules from the ocean floor comprises (1) an endless chain of conveyor sections, each section including a load chamber and a gas-accommodating float chamber; (2) means for charging gathered nodules into each successive section, followed by filling the float chamber with a gas, and (3) at the surface vessel, means for discharging both the mineral load and the buoyant gas from each successive section. This apparatus is said to be structurally simple and adapted to high capacity operations. S.J. Istoshin, G.M. Lezgintsev, M.A. Belyavsky, E.A. Kontar, and N.N. Koptyazhin, assigned to Vsesojuzny Nauchno-Issledovatelsy i Proektny Institut Zolotodobyvajuschei Promyshlennosti "Vniiprozoloto". Sept. 26, 1979. 6 pp. (B65G 15/08). Same: **Canadian 1,075,729**, dated Apr. 15, 1980.

US 4,195,426 — Manganese nodules are recovered from relatively great depths on the ocean floor using a self-propelled remote-controlled vehicle having a V-shaped tubular structure with a system of crawlers at its apexes. Underneath each side of the V-shaped structure is a scraping system for gathering nodules and directing them into a tubular conveying system for elevation to the surface ship. A given area of seabed is fully mined in a single pass of the vehicle. V. Banzoli, V. Di Tella, and P. Vielmo, assigned to Tecnomare, SPA. Apr. 1, 1980. 5 pp. (37-60). Priority: Italy, 4-1-77 (84119A/77). Same: **Canadian 1,091,259**, dated Dec. 9, 1980.

US 4,196,531 — Gathering and raising manganese nodules from the ocean floor. A cable is connected at one end to a float and at the other to collecting vehicles which land on the ocean floor at points predetermined by the landing of a ballast provided with a pile anchored in the seabed. Each collecting vehicle is moved by upward motion of means detached from the ballast. P.L. Balligand, P. Biancale, J.-P. Jacquemin, and M. Rommens, assigned to Commissariat a l'Energie Atomique. Apr. 8, 1980. 24 pp. (37-54).

Canadian, 1,082,795 — Geophysical exploration of the seabed for deposits of manganese nodules. Two partially overlapping sonar acoustic beams of different frequencies directed bilaterally away from each other are directed downward to intersect the seabed. Back-scattered and reflected echoes from the beams are transmitted and recovered with transducers on an underwater vehicle. Back-scattered vibrations are converted to electric signals indicative of back-scattered vibration amplitudes. Difference-frequency vibrations are converted to electrical signals indicative of difference-frequency vibration amplitudes. The two sets of signals are compared and interpreted. J.G. Kosalos and R.W. Cooke, assigned to Inco Ltd. July 29, 1980. 22 pp. (349-9). Priority: 11-30-76 (746,184). Same: **US 4,075,599,** dated Feb. 21, 1978 and **British 1,654,377,** dated Oct. 17, 1979.

US 4,226,035 — A one-ship system for recovering manganese nodules from the ocean floor. Plural pairs of endless ropes are moved downward over plural pairs of pulleys at the fore end of the ship and upward over plural pairs of pulleys at the aft end of the ship. Scraper buckets attached to the ropes are dragged along the ocean floor and raised and dumped when filled into the ship's hold, then recirculated to the ocean floor. N. Saito, Oct. 7, 1980. 10 pp. (37-69). Priority: Japan, 10-25-77 (52,127,755).

US 4,231,171 — Manganese nodules are recovered from the ocean floor by means of a plurality of self-propelled vehicles which move upwards and downwards between the ocean floor and a surface platform under the action of excess ballast which is partially and progressively released as the vehicle approaches the ocean floor. The vehicles are propelled along the ocean floor to gather nodules via two or more supporting units having at least one propulsion fin. Loaded vehicles are buoyed to the surface platform for docking, unloading, and energy recharging. P. Balligand, Y. Corfa, P. Lemercier, P. Marchal, and J. Vertut, assigned to Commissariat a l'Energie Atomique. Nov. 4, 1980. 14 pp. (37-54). Priority: France, 1-18-77 (77/02,188). Same: **Canadian 1,089,500,** dated Nov. 11, 1980.

US 4,232,903 — This system for recovering manganese nodules from the ocean floor at depths in the range of 4300 to 5500 m (14,000 to 18,000 ft) comprises a surface processing and control ship, an ocean bottom sub-system which includes a manoeuverable mining vehicle, and an underwater sub-system which includes a buffer means for temporarily storing nodules and for isolating the mining vehicle from the dynamics of the pipe string extending down from the surface ship. The mining vehicle is located and operated via a system of sensors and controls that defines the topography of the ocean floor and displays the movement of the vehicle and pickup and handling of the nodules. C.G. Welling, G.H. Davenport, G. Reichert, C.M. Snyder, M.C. Harrold, S.H. Donze, and F.R. Larsen, assigned to Lockheed Missiles & Space Co., Inc. Nov. 11, 1980. 23 pp. (299-4).

Suppliers of Seabed Mining Technology

In identifying the suppliers in the market, it is useful to examine the activites in the development process of seabed mining technology and the nature of participation of various parties in those activities, which determines the ownership of technology. The mining system developer engages in design activities and research and development activities for the overall system. He also engages in design activities and research and development activities for unique components which are new and do not exist elsewhere. There may also be particular components which exist elsewhere but have to be modified for the purpose of the mining system developer. In these cases, he designs the modifications and gives the specifications for modification to the existing suppliers of components. He may also engage in testing the modified component for performance. There may also be standard components which the system developer can use off-the-shelf of the existing suppliers of these components. Finally, the system developer has to integrate the new, modified and standard components into the overall mining system for operation.

Within each group of system developers, there may be division of labour among the members of the group regarding design and R & D activities of the sub-systems and components. In these cases, there may be allocation of proprietary interests between the group, on one hand, and each member, on the other. It is not quite clear to what extent the group itself and the member concerned can be considered a supplier of the particular technology. There may also be corporate relationship between the system developer and new or modified component fabricator.

Fabrication of the newly developed components may be done either by the system developer himself, or an outside fabricator may be contracted who may use his own proprietary processes. The fabricator may also sub-contract another tier of fabricators for parts, who in turn may use their own proprietary processes in fabricating the parts. The same is true about the fabrication of modified standard components.

There can be then, five categories of suppliers of technology:

system developers (with proprietary interest in system design, specification and integration; new components; possibly modified standard components);

new component fabricators (with possible proprietary interest in fabrication processes);

modified standard component fabricators (with possible proprietary interest in fabrication processes);

fabricators of parts of new or modified standard components (with possible proprietary interest in fabrication processes of these parts);

standard component suppliers (with possible proprietary interest in fabrication processes).

There seems to be agreement that out of the five categories of suppliers identified above, the standard component suppliers operate in a relatively open market. It may be pointed out that this category is not "tied down" with a particular nodule mining technology developer; in fact, it can be a supplier of technology to any nodule mining technology developer. Currently, the members of this category supply technologies to other industries as well, offshore oil and gas, for example. Manufacturers of pipes, pumps, navigation equipment and computers fall in this category. There is considerable controversy regarding the openness of market where the other four categories, especially the first category, system developers, participate. Table B1 lists the currently active system developers.

However, possibilities exist that technology can be acquired by launching a design and R & D programme. The fact that several groups of firms have embarked on their own design, research and development of technology independently of one another, and without directly relying on one another have proven, to a greater or lesser degree, the technical feasibility of their own mining systems, implies that nodule mining technology is not a "secret" or "mystery" known to a single entity. It can be mentioned that enough technical capabilities exist outside the currently engaged groups which can be pooled together to acquire nodule mining technology without being dependent on them. It is also noteworthy that for various reasons, a relatively long gap occurred between the pilot phase and the large-scale demonstration phase; some of the R & D teams of the pioneer groups have been dismantled and members of these teams can be considered "footloose" in the market. The experience and expertise of these personnel can be tapped. Table B2 presents a list of enterprises that are considered capable of assisting in a design and R & D programme for nodule mining.

Attempts have been made in identifying the components of exploration, mining and ore transfer technologies which cannot be acquired off-the-shelf of existing suppliers and need to be newly innovated or modified. It appears that *substantial new innovation* is required for the following components:

Collector sub-system:	Nodule pick up
	Collector propulsion
Mining ship sub-system:	Ship/ship connection
Ore transfer:	Reslurry equipment

Substantial modification is required in existing components for the following:

Lift sub-system:	Connectors
	Hoses
	Power cable
	Instrumentation
Mining ship sub-system:	Mining system control

Table B3 presents a list of component suppliers for exploration, mining and ore transfer technologies following the component breakdown of Figs 4, 11 and 19. The list has been compiled from trade journals and directories. The list is intended to be illustrative and in no way exhaustive. It should also be mentioned that owing to the very nature of the compilation exercise, developing countries and Eastern Europe (Socialist) countries are under-represented in the list.

Table B1. List of Enterprises Currently Engaged in Mining System Development

1. Kennecott Consortium (KCON)

Year of formation	Country of registration	Head office	Service contractor
January 1974	Unincorporated		

COMPOSITION OF CONSORTIUM

Participants	Parent company	Country of origin of parent company
Kennecott Corporation	Sohio	United States
RTZ Deepsea Enterprises Ltd.	Rio Tinto-Zinc Corporation Ltd.	United Kingdom
Consolidated Gold Fields, PLC	Same	United Kingdom
BP Petroleum Development Ltd.	British Petroleum Company, Ltd.	United Kingdom
Noranda Exploration, Inc.	Noranda Mines, Ltd.	Canada
Mitsubishi Group	Mitsubishi Corporation Mitsubishi Metal Corporation Mitsubishi Heavy Industries, Ltd.	Japan

Table B1 (continued)

2. Ocean Mining Associates (OMA)

Year of formation	Country of registration	Head office	Service contractor
May 1974	Partnership registered in Virginia, United States	Gloucester Point, Virginia	Deepsea Ventures, Inc. (Gloucester Point, Virginia)

COMPOSITION OF CONSORTIUM

Participants	Parent company	Country of origin of parent company
Essex Minerals Company	United States Steel Corporation	United States
Union Seas, Inc.	Union Minière S.A.	Belgium
Sun Ocean Ventures	Sun Company, Inc.	United States
Samin Ocean, Inc.	Ente Nazionale Idrocarburi (ENI)	Italy

3. Ocean Management Incorporated

Year of formation	Country of registration	Head office	Service contractor
February 1975	Incorporated in the United States	Administrative office in New York, New York	

COMPOSITION OF CONSORTIUM

Participants	Parent company	Country of origin of parent company	
Inco, Ltd.	Same	Canada	
AMR (Arbeitsgemeinschaft Meerestechnisch-gewinnbare Rohstoffe)	Metallgesellschaft AG		
	Preussag AG	Federal Republic of Germany	
	Salzgitter AG		
SEDCO, Inc.	Same	United States	
Deep Ocean Mining Company, Ltd. (DOMCO)	23 companies	Japan	

Table B1 (continued)

4. Ocean Minerals Company (OMCO)

Year of formation	Country of registration	Head office	Service contractor
November 1977	United States partnership	Mountain View, California	

COMPOSITION OF CONSORTIUM

Participants	Parent company	Country of origin of parent company
Amoco Ocean Minerals Company	Standard Oil of Indiana	United States
Lockheed Systems Company, Inc.	Lockheed Aircraft Corporation	United States
	Lockheed Missiles and Space Company, Inc. (subsidiary of Lockheed Aircraft Corporation)	United States
Ocean Minerals, Inc.	Billiton B.V. (a Netherlands company of the Royal Dutch/Shell group)	Netherlands

BKW Ocean Minerals BV
(a Netherlands
subsidiary of the Royal
Bos Kalis Westimster
Group, NV)

United States

Netherlands

5. Association Française pour l'Étude et la Recherche des Nodules (AFERNOD)

Year of formation	Country of registration	Head office	Service contractor
1974	France	Paris	

COMPOSITION OF CONSORTIUM

Participants
Centre National pour l'Exploitation des Océans (CNEXO)

Commissariat à l'Energie Atomique (CEA)

Société Métallurgique le Nickel (SLN)

Chantiers de France-Dunkerque

Table B1 (continued)

6. Deep Ocean Resources Development (DORD) Company, Ltd, formerly Deep Ocean Minerals Association (DOMA)

Year of formation	Country of registration	Head office	Service contractor
March 1974	Japan, as a public corporation	Tokyo	

COMPOSITION OF ASSOCIATION

Members		Members
C. Itoh and Company, Ltd.		Hitachi Shipbuilding and Engineering Company, Ltd.
Marubeni Corporation		Ishikawajima-Harima Heavy Industries Company, Ltd.
Mitsubishi Corporation		
Mitsui and Company, Ltd.		Kawasaki Heavy Industries Company, Ltd.
Nichimen Company, Ltd.		Mitsubishi Heavy Industries, Ltd.
Nissho Iwai Corporation		Mitsui Engineering and Shipbuilding Company, Ltd.
Sumitomo Corporation		Nippon Kokan K.K.
		Sumitomo Heavy Industries, Ltd.
Dowa Mining Company, Ltd.		Ebara Corporation
Furukawa Company, Ltd.		Meidensha Manufacturing Corporation
Japan Metals and Chemicals Company, Ltd.		
Mitsubishi Metal Corporation		
Mitsui Mining and Smelting Company, Ltd.		
Nippon Mining Company, Ltd.		
Nippon Yakin Kogyo Company, Ltd.		Kawasaki Steel Corporation
Nittetsu Mining Company, Ltd.		Kobe Steel, Ltd.

Pacific Metals Company, Ltd.
Sumitomo Metal Mining Company, Ltd.

Nippon Steel Corporation
Sumitomo Metal Industries, Ltd.

Iino Kaium Kaisha, Ltd.
Mitsui O.S.K. Lines, Ltd.
Nippon Yusen K.K.

The Fujikura Cable Works, Ltd.
Sumitomo Electric Industries, Ltd.

Nippon Electric Company, Ltd.
Victor Company of Japan, Ltd.

Kyokuyo Company, Ltd.

7. Public Enterprise of the Union of Soviet Socialist Republics
8. Public Enterprise of India
9. Public Enterprise of People's Republic of China

Table B2. List of Enterprises Potentially Capable of Mining System Development

Enterprise	Country	Enterprise	Country
Petrobras	Brazil	Stolt Nielsen Seaway	Norway
Tri-Ocean Engineering	Canada	Hyundai Construction	Republic of Korea
CFE	France	Sembawang Engineering	Singapore
ETPM	France	Swedyards	Sweden
MATRA	France	Press Group	United Kingdom
Thyssen	Germany (Federal Republic of)	Seaforth Maritime	United Kingdom
SAIPEM	Italy	Vickers Offshore	United Kingdom
Protexa	Mexico	Brown and Root	United States
Gasunie Engineering	Netherlands	Fluor	United States
IHC Holland	Netherlands	Global Marine	United States
Main Structure Consultants	Netherlands	J. Ray McDermott	United States
Rhine-ScheldeOverolme RSV/Offshore	Netherlands	Santa Fe International	United States
Aker Group	Norway	Science Applications	United States
Kvaerner Engineering	Norway	TRW	United States

(The systems development function includes the technical and administrative tasks necessary for the design, specification, and supervision of fabrication of the overall mining system. The function requires highly skilled administrtive, engineering and scientific staff who are able to apply an integrated systems management approach to a nodule mining project.)

Table B3. List of Component Suppliers of Nodule Exploration, Mining and Ore Transfer Technologies

E.0   *Exploration System*

E.1   *Exploration Ship/Platform.* Options regarding exploration ships include: (a) lease or acquire basic platform and outfit for survey; (b) lease existing survey/research ship.

Option (a)
Australian Offshore Services (Australia); Union De Remorquage Et de Sauvetage (Belgium); Petrobras (Brazil); Federal Offshore Services (Canada); China National Machinery Export (China, People's Republic of); A.P. Moller (Maersk Supply Service) (Denmark); Bugsier Reederie u. Bergungs, Deutsche Offshore Gesellschaft, Hapag Lloyd Transport and Service, Unterweser Reederei (Germany, Federal Republic of); Terminal Installations (Liberia); International Transport Contractors, Smit International Nederland, Smit Lloyd, Wijsmuller (Netherlands); Active Marine Offshore Services, Arvid Bergvall, jr. Shipping Brodene Klovning, Buchanan Shipping London, Bugge Supply Ships, Edda Supply Ships, Granstad Supply, Karmoy Supply, Helge Kyvik, John Larsen, Norway Supply Ships, G.C. Rieber, Salvator Norsk Bjergningskompani, Sandoy Supply, Seaway Supply and Support Ships, Stad Seaforth Shipping, Stolt-Nielsen Rederi, Supply Service, Ugelstads Rederi, Viking Supply Ships, Wilhelmsen Offshore Services, Yngvar Hvistendahl (Norway); Offshore Marine, Offshore Supply Association (United Kingdom); Acadian Marine, Aqua Marine, Astro Marine, Euro-Pirates International, Gulf-Mississippi Marine, Sealcraft Operators (United States).

Option (b)
Compagnie Generale de Geophysique (France); Prakla-Seismos (Germany, Federal Republic of); Heerema, Martinus Nijhofflaanz (Netherlands); Decca Survey Systems, S and A Geophysical (United Kingdom); Alcoa Marine, Cape Fear Technical Institute, EG+G, Esso Seismic, Geophysical Services, Global Marine, Grant Geophysical, Gulf Research and Development, Hawaii Institute of Geophysics, Lamont-Doherty Geological Observatory, Marine Biomedical Institute, Mobil Oil, Petty-Ray Geophysical, Rosenstiel School of Marine and Atmospheric Sciences, Scripps Institution of Oceanography, Sealcraft Operators, Seinograph Service, Seismic Explorations, Shell Oil, State Boat, Teledyne Exploration, Texas A and M University, Tractor Marine, University of Rhode Island, University of Washington, Western Geophysical, Woods Hole Oceanographic Institute (United States).

E.2   *Exploration Equipment*

E.2.1   *Bottom Sampling Devices*

E.2.1.1   *Grab Sampler*
Hydro Bois, Hydrow (Germany, Federal Republic of); Rigosha (Japan); Bergen Nautik (Norway); Kelvin Hughes (United Kingdom); EG+G, G.M. Manufacturing, Hydro Products, Kahl (United States).

E.2.1.2   *Corer*
Askonia, Hydrow (Germany, Federal Republic of); Rigosha, T.S.K. (Japan); Mashphribor (USSR); Alpine, G.M. Manufacturing, Hydro Products, Kahl, Ocean Instrument (United States).

Table B3 (continued)

E.2.1.3	*Dredge*
	Rigosha, T.S.K. (Japan); G.M. Manufacturing, Kahl (United States).
E.2.2.	*Bottom Characteristics*
E2.2.1	*Camera,* E.2.2.2 *TV,* E.2.2.3 *Strobe*
	Serel (France); IBAK (Germany, Federal Republic of); Marconi (United Kingdom); Benthos, EDO Western, Hydroproducts, Kinergetics, OEC, Sub-sea Systems (USA).
E.2.2.4	*Echo Sounder*
	Fathom Oceanology, Huntec ('70) Mesotech (Canada); Comex Industries, Geomecanique (France); Atlas Werke, Elac, Krupp Atlas Elektronik (Germany, Federal Republic of); Nippon Electric (Japan); Simrad (Norway); Kelvin Hughes (UK); Ametek, ANG, Apelco, Bendix EG+G, EDO Western, EPC, Fisher, General Instruments, Gould, Innerspace Technology, International Submarine Technology, International Transducer, Interocean Systems, ITT Decca Marine, Klein Associates, Marconi Marine, MASSA Products, Nektron, Ocean Research Equipment, Raytheon Ocean Systems, Teledyne Geotech (USA).
E.2.2.5	*Multi-Beam Sonar*
	Scripps Institution of Oceanography (USA).
E.2.2.6	*Side-Scan Sonar*
	Atlas Werke, Elac (Germany, Federal Republic of); Simrad (Norway); Kelvin Hughes (UK); Klein Associates (USA).
E.2.3	*Surface Characteristics*
E.2.3.1	*Wave Tracker,* E.2.3.2 *Current Meter,* E.2.3.3 *Weather Station*
	Applied Microsystems (Canada); Tasurumi Precision Instrument (Japan); Aanderra Instruments (Norway); Plessey (UK); Belfort Instrument, Bendix, Braincon, EG+G Sea-Link Systems, Endeco, Hydro Products, Kahl Scientific Instrument, Leupold + Stevens Instrument, Litton Industries, Magnavox Government and Industrial Electronics, Marine Advisors, Neil Brown Instrument System, Olympic Instrument, W+L.E. Gurley (USA).
E.2.4.	*Deck Equipment*
E.2.4.1	*Winches,* E.2.4.2 *Cranes*
	Fathom Oceanology, John T. Hepburn Mechanical Division (Canada); Geisselback Elektrontechniek, Hoogerwerff Staalkabel (Netherlands); Elkem-Spiegerverket, Hydraulik Brattvaag (Norway); British Ropes (UK); Appleton Marine (Appleton Machine), Beebe Bros, Carolina Steel and Wire, Cortland Line, DCMA International, EDO Western, Environmarine Systems, Frazer, General Oceanics, Hib Cranes and Loaders, Industrial Electric Reels, Interocean Systems, Macwhyte Wire Rope, Marine Crane Otis Engineering, Preformed Line Products Sea Mac, Superior Switchboard and Devices, United States Steel, Wall Rope Works, Zippertubing (USA).
E.2.4.3	*Photography Laboratory*
	Geostructures (USA).
E.2.4.4	*Instrument Vans*
E.2.4.5	*Rigging*
	AMCON, American Hoist, Bethlehem Steel (USA).

E.3	*Data Evaluation*
	Geonautics (Canada); Gardline Hydrographic Surveys, Hydrotask, Offshore Industries, Tetra Tech, Tracor Marine (USA).
E.3.1	*Geology Laboratory*
	Geostructures (USA)
E.3.2	*Data Management System*
	Hydroproducts, Interstate Electronics, Metrox, Science Engineering Associates (USA).
E.3.3	*Software*
	Hydroproducts, Interstate Electronics, Metrox, Science Engineering Associates (USA).
E.3.4	*On Board Computer*
	MOM (UK); Digital Equip, Frieden, Hewlett Packard, Wang (USA).
M.0	*Mining System*
M.1	*Collector*
M.1.1	*Collector Structure*
M.1.2	*Collector Deployment*
M.1.3	*Nodule Pick Up*
M.1.4	*Nodule Oversize Eliminator*
M.1.5	*Sediment Eliminator*
M.1.6	*Motors*
	Pleuger (Germany, Federal Republic of); Sulzer (Switzerland); Franklin, General Electric, Hansome, Louis Allis (USA).
M.1.7	*Collector Pumps*
	Plueger (Germany, Federal Republic of); Hydrotask (USA).
M.1.8	*Steering*
M.1.9	*Propulsion*
M.1.10	*Collector Valves*
M.1.11	*Collector Buoyancy*
M.1.12	*Cables and Connectors*
M.1.13	*Collector Instrumenation*
	Atlas Werke, Elac, IBAK, Krupp Atlas Electronik (Germany, Federal Republic of); Nippon Elec. (Japan); Simrad (Norway); Kelvin Hughes (UK); EDO Western, Hydroproducts, Klein Associates (USA).
M.2	*Lift Subsystem*
M.2.1	*Lift Pumps*
	Sulzer (Switzerland); Simon-Warman, Weir (UK); Byron Jackson, Georgia Iron Works, Mobil Pully and Machine Works, Morris Pumps, Thomas Foundaries, Worthington (USA).
M.2.2	*Lift Motors*
	KSB, Plueger (Germany, Federal Republic of); Sulzer (Switzerland); General Electric (Argo), Hayward Tyler, Louis Allis (USA).
M.2.3	*Lift Compressors*
	Mitsubishi (Japan); Allis Chalmers, Ingersol Rand, Schramm (USA).
M.2.4	*Lift Subsystem Pipe,* M.2.5 *Lift Subsystem Pipe Joints*
	Vallourec (France); Mannesman Rohren-Werke (Germany, Federal Republic of); Japan Steel, Sumitomo Steel (Japan); British Steel (UK); Armco Steel, Bethlehem Steel, Cameron Iron Works, Jorgensen Steel, Sandusky Steel, U.S. Steel (USA).
M.2.6	*Fairings*

Table B3 (continued)

	Fathom (Canada).
M.2.7	*Valves*
M.2.8	*Power Cables,* M.2.9 *Connectors*
	Industrie Pierelli (Italy); Boston Insulated Wire and Cable, Burton Electrical Engineering, Cortland Line, Crouse-Hinds, Electro Oceanics, Environmarine System, Kintek, MASSA Products, Poly Scientific, Rochester, Simplex Wire and Cable, United States Steel, Vector Cable, Viking Connectors (USA).
M.2.10	
M.3.2.3	*Hose (Slurry)*
T.2.2	Dunlop Industrial (Australia); Coflexip (France); Uniroyal (Italy); Bridgestone, Yokohama (Japan); Vredestein (Netherlands); Dunlop (UK); Gates, Goodrich, Goodyear, Simplex Wire and Cable (USA).
M.2.11	*Lift System Instrumentation*
M.3	*Surface Sub-system*
M.3.1	*Subsea Equipment Handling*
M.3.1.1	*Pipe Handling* M.3.1.2 *Collector Handling,* M.3.1.3 *Heavy Lift System.* M.3.1.4 *Fairing Handling,* M.3.1.5 *Gimbal Platform,* M.3.1.8 *Power Cable Handling.*
	Terminal E. Emshaven (Netherlands); Brown and Root, Byron Jackson, J. Ray McDermott, Lockheed, Santa Fe International, Varco International, Western Gear (USA).
M.3.1.6	*Gimbal Bearings*
	Fag Bearings (Germany, Federal Republic of); Garlock Bearing, Lockheed, Lord Kinematics, Merriman Bearing, Oil States Rubber, SKF Bearings (USA).
M.3.1.7	*Heave Compensator*
	Hydra-Rig, Regan Forge, Rucker, Vetco, Western Gear (USA).
M.3.1.9	*Derrick*
	Continental Emsco, Lee C. Moore, Pyramid Derrik, Sterns-Roger, Western Gear (USA).
M.3.2	*Material Handling*
M.3.2.1	
T.2.1	*Pumps*
	Kelly and Lewis Pumps (Australia); Holthuis (Netherlands); KSB (Germany, Federal Republic of); Bowie Industries, Simon Warman, Thomas Foundries (USA).
M.3.2.2	*Centrifuges, Concentrators*
	F.L. Smidth (Denmark); KHD Humbolt/Nedag (Germany, Federal Republic of); Skega (Sweden); Liquisolid (UK); Centrifugal and Mechanical Industries, Deister Concentrator, Krebs, Linatex, Townley Eng and Mfg. (USA).
M.3.2.4	*Air/Water Separator* (Air Lift System Only)
M.3.2.5	*Grizzley, Feeders, Conveyors*
	Merer, Ross Eng, Stracham and Henshaw (UK); Baber Greene, Bucyrus-Erie, Cable Belt, Portec, Ramsey Engineering, Rexnord (USA).
M.3.2.6	*Screens*
	Locker Ind. (Australia); W.S. Tyler (Canada); Skega (Sweden); Rexnord (USA).

M.3.2.7	*Effluent Discharge System*
M.3.2.8	*Buffer Storage*
	Marcona (USA)
M3.2.9	Ship/Ship Connection (Material Transfer)
	PHB Engineering (Australia); Mac Gregor-Comarain (USA).
M.3.	*Surface Sub-system*
M.3.3	*Platform*
M.3.3.1 T.1.1	*Ship Hull*

(Note: Only representative shipyards in developing countries capable of and experienced in, building large ships are listed. In every major developed country, there are many shipyards able to meet the requirements). Estado S.A. Astilleros Y Fabricas Navales (Argentina); Arab Shipbuilding and Repair (Bahrain); Comercio E Navegacao, EMAQ-Engenharia E Maquinas (Brazil); Cochin Shipyard (India); Malaysia Shipyard and Engineering (Malaysia); Karachi Shipyard and Engineering Works (Pakistan); Atlantic, Gulf and Pacific Co. of Manila (Philippines); Hyundai Shipbuilding and Heaving Industry, Pae Sun Shipbuilding and Engineering, Sansung Shipbuilding (Korea, Republic of); Asia-Pacific Shipyard, Bethlehem Singapore, Far East Levington Shipbuilding, Keppel Shipyard, Marathon Le Tournenn (Singapore).

M.3.2.2 T.1.2	*Main Propulsion*

Escher Wyss, Lohmann and Stolterfoht, Reintjes, Voith-Schneider (Germany, Federal Republic of); Lips Propellor Works, Schottel-Werft (Netherlands); Duramax International, Newage Engineers (UK); Armco Steel, Columbian Bronze, Coolidge Propellors, Farrel, Ferguson, Kahlenberg Brothers, Philadelphia Gear, Marine Gears, Michigan Wheel, Waukesha Bearing (USA).

M.3.3.3 T.1.3	*Thrusters*

Maritime Industries (Canada); Schottel (Germany, Federal Republic of); Elliot, Kamenwa (Bird Johnson), Omnithruster, PSI Propulsion System (USA).

M.3.3.3 T.1.4	*Steering*

Hitachi Zosen (Japan); Stork-Services (Netherlands); DeLaval, Jered Industries, Paul Munroe Marine Division, Penco, Sperry Marine Systems, Torrington Bearing (USA).

M.3.3.5 T.1.5	*Navigation*

Canadian Marconi (Canada); Sercel (France); Krupp Atlas-Elektronik (Germany, Federal Republic of); Ametek, Edo Western, Digital Marine Electronics, Epsco, Magnavox Government and Industrial Electronics, Micro Marine, Navidyne, Navigation Communication Systems, Raytheon Marine, Satellite Positioning Systems, Sperry Marine Systems, Tracor (USA).

M.3.3.6	*Power Generation*
	Rolls Royce (UK); General Electric, Pratt and Whitney, SOLAR (USA).
M.3.3.7	*Deck Cranes and Winches*
	A.B. Hagglund and Soner, (Sweden); J.H. Fenner (UK); Amcom, Bucyrus-Erie, Hydranautics, Marathon Letourneau, Paceco, Pedestal Crane, Skagit, Unit Mariner, Western Gear (USA).

Table B3 (continued)

M.3.3.8 T.1.7	*Anchor Handling* Lake Shore, Skagit, Western Gear (USA).
M.3.3.9	*Underwater Navigation* Mesotech Systems (Canada); Oceano-Instruments, Sercel, Thomson (France); Krupp Atlas Elektronik (Germany, Federal Republic of); TSK (Japan); Kongsberg, Simrad Offshore Division (Norway); Marconi International Marine Company (UK); Benthos, California Computer Products, Cubic Western Data, Del Norte, Edo Western, Environmental Devices, Environmarine Systems, EPSCO Marine, Hewlet Packard, Honeywell, Houston Instruments, Motorola Position Determining Systems, Ocean Research Equipment, Ocean/Seismic/Survey, Offshore Navigation, Sea-Link Systems, Sonotech, Tektronix, Teledyne Hastings-Radist, Varian Graphics, Versatec, Zeta Research (USA).
M.3.3.10	*Moonpool Closure*
M.3.3.11	*Mining System Control* Cubic, Foxboro Instruments, General Dynamics, Honeywell, IBM, Litton Industries (USA).
M.3.3.12 T.1.9	*Hotel Services* Aqua Chem, Argo Marine, Bailey Jomier, Custom Alloy, Demco, Envirovac, Fast Sewage System, IDT, Kelvinator, Marine Moisture Control, Marland Environmental System, Oceanic Elec. Mfg., Oreck, Perko, Phoenix Products, Red Fox Industries, Robvon Racking Ring, Sigma Treatment Systems, Weir Pumps (USA).

Appendix C

Bibliography

A. GENERAL

(Includes discussion of technology)

Albers, J. P., 1974. Some geologic and resource aspects of ferromanganese nodules. *Colloque International sur l'Exploitation des Oceans*, 2ème, Bordeaux.

Amann, H. M., 1975. Definition of an ocean mining site, Offshore Technology Conference. *Proceedings*, OTC, Dallas, Texas, vol. 1, pp. 891–908.

Amos, A. F. *et al.*, 1972. Effects of surface-discharged deep sea mining effluent. *Journal of Marine Technology Society*, **6**, no. 4.

Andrews, J. 1974. Recent investigations of the geological setting of manganese nodule deposits, and new approaches to the problem. *Colloque International sur l'Exploitation des Oceans*, 2ème, Bordeaux.

Anon., 1976a. Summa Corp.'s mining plans firm despite spectacular publicity. *Mining Engineering*, pp. 6, 12, 15–20.

Anon., 1976b. *Definition of Mining System Parameters of Importance to an Environmental Study of Deep Ocean Mining*, National Oceanic and Atmospheric Administration, U.S. Department of Commerce, NOAA/ DOMES Advisory Panel Meeting, 24 February.

Anon. 1977a. Converted drillship will begin deepsea mining test in October. *Ocean Industry*, **12**, no. 6.

Anon., 1977b. Mining the seabed. *Mechanical Engineering*, February, pp. 88–89.

Anon., 1977c. Who owns the oceans. *Compressed Air*, June, pp. 10–12.

Anon., 1978a. Manganese ore taken from bed of Pacific. *Lloyds List*, (No. 50921).

Anon., 1978b. Pacific hunt for vast manganese deposits. *Lloyds List*, (No. 50868).

Archer, A. A., 1976. Prospects for the exploitation of manganese nodules: the main technical, economic and legal problems. *CCOP/SOPAC Technical Bulletin* No. 2.

Arthur D. Little, Inc., 1977. *Technical and Economic Assessment of Manganese Nodule Mining and Processing*, United States Department of Interior, 92 pp. (U.S. Dept. of Interior Contract No. 14-01-0001-2114).

Arthur D. Little, Inc., 1978. *Technical and Economic Assessment of Manganese Nodule Mining*, Office of Minerals Policy Research and Analysis, U.S. Department of the Interior, Washington, D.C.

Bailly, P. A., 1976. The problems of converting resources to reserves. *Mining Engineering*, January.

Bastien-Thiry, H., 1977. Prospecting—evaluation, Paper presented at the Seminar on Exploitation of Deep Seabed, Brussels, Unpublished manuscript (Report No. 7).

Bastien-Thiry, H., J.-P. Lenoble and P. Rogel, 1977. French exploration seeks to define mineable nodule tonnages on Pacific floor. *Engineering and Mining Journal*, **178**, no. 7.

Battelle Memorial Institute, 1971. *Environmental Disturbances of Concern to Marine Mining Research, A Selected Annotated Bibliography*, National Oceanic and Atmospheric Administration, U.S. Department of Commerce, Rockville, Maryland, (NOAA Technical Memorandum ERL MMTC-3).

B.D.M. Corp., 1977a. *A Technology Assessment of Offshore Industry and Its Impact on the Maritime Industry 1976–2000*, U.S. Department of Commerce, Washington, D.C., (U.S. Dept. of Commerce, Office of Commercial Development Report No. MA-RD-940-78003, BDM Report No. BDM/W-77-425-TR).

B.D.M. Corp., 1977b. *A Technology Assessment of Offshore Industry — Current Status, Trends and Forecast 1976–2000*, U.S. Department of Commerce, Washington, D.C., (U.S. Dept. of Commerce, Office of Commercial Development Report No. MA-RD-940-78002, BDM Report No. BDM/W-77-425-TR).

Bhatt, J. J., 1979. *Applied Oceanography: Mining, Energy and Management*, University Microfilms International, Ann Arbor, Michigan, 245 pp.

Bischoff, J. L. and D. Z. Piper, eds., 1979. *Marine Geology and Oceanography of the Pacific Manganese Nodule Province*, Plenum Press, New York, 839 pp.

Bischoff, J. L. and R. J. Rosenbauer, 1976. *Recent Metalliferous Sediment in the Pacific Manganese Nodule Area. Geology and Geochemistry of Site C, Deep Ocean Mining Environmental Study, NE Pacific Nodule Province*, U.S. Geological Survey, Menlo Park, California (U.S.G.S. Open File Report No. 76-548).

Boin, U., 1980. Limits and possibilities of deep sea mining for the extraction of mineral raw materials — the case of manganese nodules, *Mining Magazine*, January.

Booda, L. L., 1976. Marine mining faces bright economic future, *Sea Technology*, August.

Borgese, E. M., 1982. *Ocean Mining and Developing Countries: An Approach to Technological Disaggregation*, United Nations Industrial Development Organization, Vienna, Austria. (Document UNIDO/IS.345 of 5 October).

Bureau de Recherche Geologique et Mineralogique, 1978. *Resources Minerale sous-Marines Comptes Rendus, Proceedings of Conference,* Orleans, France, 23–27 October.

Burns, R. E. *et al.,* 1978. *Observations and Measurements during the Monitoring of Deep Ocean Manganese Nodule Mining Tests in the North Pacific, April–May 1978,* National Oceanic and Atmospheric Administration, U.S. Department of Commerce, Rockville, Maryland. (Advance Draft of a NOAA Report on OMI's Pilot Mining Test during DOMES Phase II).

Burns, R. G., ed., 1979. *Marine Minerals,* Bookcrafters Inc., Chelsea, Michigan.

Caldwell, A. B., 1971. Deepsea Ventures readying its attack on Pacific nodules. *Society of Mining Engineers,* October.

Clauss, G., 1972. Theoretical and experimental investigations of deep ocean mining systems and their economic implications, *Second International Ocean Development Conference,* Tokyo, Japan, 5–7 October.

Clay, C. S. and H. Medwin, 1977. *Acoustical Oceanography: Principles and Applications,* John Wiley and Sons, New York.

Cozens, A., 1978. Ocean mining: the treasure hunt begins. *Offshore,* **38,** no. 2, pp. 154–162.

Cronan, D. S., 1976. Mineral exploration in the Indian Ocean, *Ocean Industry,* **11,** no. 8, pp. 86–87.

Cronan, D. S., 1980. *Underwater Minerals,* Academic Press, London, 362 pp.

Cruickshank, M. J., 1974. Environmental impact analysis for marine mining operations. *Colloque International sur l'Exploitation des Oceans,* 2ème, Bordeaux, France.

Dames and Moore and EIC Corporation, 1977. *Description of Manganese Nodule Activities for Environmental Studies: Vol. II — Transportation and Waste Disposal Systems,* National Oceanic and Atmospheric Administration, U.S. Department of Commerce, Rockville, Maryland. (NTIS No. PB 274 914).

Derkman, K. J., R. Fellerer, and H. Richter, 1981. Ten years of German exploration activities in the field of marine raw materials. *Ocean Management,* **7,** pp. 1–8.

Dunne, J. A., 1976. The Seasat-A project: an overview, Marine Technology Society — Institute of Electrical and Electronic Engineering, *Oceans '76,* pp. 10A-1–10A-5.

Enzer, H., 1979. Ocean mining, an assessment of current technology and its economics. *10th World Mining Congress,* September 1979, Istanbul, Turkey, vol. 4, no. 13, pp.1–20.

European Economic Community, 1977. Papers Presented at the Seminar on the Exploitation of the Deep-Seabed, under the Auspices of European Economic Community for the Benefit of the ACP Experts to the UN Conference on the Law of the Sea.

114   Manganese Nodule Exploration and Mining Technology

Felix, D., 1980. Some problems in making nodule abundance estimates from sea floor photographs, *Marine Mining*, **2**, no. 3.

Fewkes, R. H. *et al.*, 1979. *Development of a Reliable Method for Evaluation of Deep Sea Manganese Nodule Deposits*, U.S. Bureau of Mines, (USBM Final Report, Contract No. GO 274013-MAS).

Fewkes, R. H. *et al.* 1980. *Evaluation of Metal Resources at and near Proposed Deep Sea Mine Sites*, U.S. Dept. of Commerce, Washington, D.C., 15 Feb. 1980, 242 pp. (NTIS-PB80-228992).

Flipse, J. E., 1969. Developments in ocean exploration and mining, *American Mining Congress Convention*, San Francisco, California, 12–22 October.

Flipse, J. E., ed., 1979. *Deep Ocean Mining*, The American Society of Mechanical Engineers, New York.

Flipse, J. E., M. A. Dubs and B. J. Greenwald, 1973. Pre-production manganese nodule mining activities and requirements, in United States Senate, 93rd Congress, 1st Session, *Mineral Resources of the Deep Seabed: Hearings before the Subcommittee on Minerals, Materials and Fuels of the Committee on Interior and Insular Affairs on S.1134*, U.S. Government Printing Office, Washington, D.C. pp. 602–700.

Gauthier, M. and C. Charles, 1979. French technological development in nodule mining. Offshore Technology Conference, *1979 Proceedings*, OTC, Dallas, pp. 101–105.

Geological Survey of Japan, 1979. *Deep Sea Mineral Resources Investigation in the Central Western Part of Central Pacific Basin*, GSJ, Hisamoto, Takatsu-ku, Kawasaki-shi, Japan.

Glasby, G. P., 1974. Exploitation of manganese nodules in South Pacific, *Papers from Conference on Circum-Pacific Energy and Mineral Resources*, 26–30 August, Honolulu, Hawaii, pp. 386–389.

Glasby, G. P., ed., 1977. *Marine Manganese Deposits*, Elsevier Publishing Co., Amsterdam, Holland.

Glasby, G. P., 1982. Marine mining and mineral research activities in Europe. *Marine Mining*, **3**, nos. 3–4, pp. 379–409.

Glasby, G. P., 1983. The three-million-tons per year manganese nodule mine site: an optimistic assumption? *Marine Mining*, **4**, no. 1, pp. 73–77.

Glasby, G. P. and H. R. Katz, eds., 1976. *Marine Geological Investigations in the Southwest Pacific and Adjacent Areas*, CCOP/SOPAC Technical Bulletin, No. 2.

Govier, G. W., and K. Aziz, 1972. *The Flow of Complex Mixtures*, Van Nostrand Reinhold Company.

Greenfield, W. D. and A. Lubinski, 1967. Use of bumper subs when drilling from drilling vessels. *American Institute of Mining Engineering (AIME) Petroleum Transactions*, December.

Greenslate, J., 1976. The IODE/NSF manganese nodule project: a review of progress, Marine Technology Society — Institute of Electrical and Electronic Engineering. *Oceans '76*, pp. 2D-1–2D-9.

Greenslate, J., 1980. Manganese concentration wet density: a marine geochemistry constant, *Marine Mining*, 1, p. 125.

Grote, P. B. and W. A. Coleman, 1978. *A Preliminary Study of the Consequences of Deep Ocean Mining Technology Transfer*, Science Applications, Inc., La Jolla, California (SAI Report No. SAI-78-1026-LJ, Prepared for U.S. Department of the Interior).

Grote, P. B. and W. Gayman, 1979. *A Technical Basis for the Development of Deep Ocean Mining Regulations*, Science Applications, Inc., La Jolla, California (SAI Report No. SAI-057-79-876-LJ, Prepared for U.S. Bureau of Mines).

Halbach, P. and P. Winter, 1982. *Marine Mineral Deposits — New Research Results and Economic Prospects*, Proceedings of Clausthaler Workshop, September.

Halkyard, J. E., 1979. Deep ocean mining — current status and future prospects. *Ocean Industry*, 14, no. 5, pp. 49–51.

Holser, A. F., 1976. *Manganese Nodule Resources and Mine-site Availability*, Ocean Mining Administration, Department of the Interior, Washington, D.C. (Professional Staff Study).

Horn, D. R., ed., 1972. *Ferromanganese Deposits in the Sea-Floor*, Columbia University Press, New York, 293 pp.

Horn, D. R., B. M. Horn and M. N. Delach, 1973a. *Factors which Control the Distribution of Ferromanganese Nodules and Proposed Research Vessel's Track, North Pacific Phase II Ferromanganese Program NSF/IDOE*, Office for the International Decade of Ocean Exploration, Washington, D.C. (Technical Report No. 8, NSF GX 33616).

Horn, D. R., B. M. Horn and M. N. Delach, 1973b. *Ocean Manganese Nodules Metal Values and Mining Sites*, Office for the International Decade of Ocean Exploration, National Science Foundation, Washington, D.C. (Technical Report No. 4).

Igrevsky, V. I. *et al.*, 1972. Main task for marine exploration geology in USSR. *Sovietskya Geologiya*, pp. 9–24.

Joyce, C., 1978. Seabed mining turns up few environmental problems. *New Scientist*, p. 733.

Kaufman, R., 1974. The selection and sizing of tracts comprising a manganese nodule ore body, in Offshore Technology Conference, *Preprints*, OTC, Dallas, Texas, vol. II, pp. 283–293.

Kaufman, R., 1976. Ocean module mining—progress and problems, in McLeod, C. R., and Pasho, D. W., eds., Papers presented at the Seventh Underwater Mining Institute, *op. cit.*

Kaufman, R., 1980. Deep ocean mining — 1980 status report. Institute of Electrical and Electronic Engineers, *Proceedings Oceans '80*.

Kaufman, R. and R. Greenwald, 1972. Manganese nodule mining — technical progress and jurisdictional uncertainty. *Oceanology 72*, Brighton, U.K.

Kent, P., 1980. *Minerals from the Marine Environment*, Edward Arnold Publishers Ltd., London, 88 pp.

Kerr, R. A., 1978. Glomar Explorer: new era in deep-sea drilling? *Science*, **200**, pp. 1254–1255.

Kildow, J., ed., 1980. *Deepsea Mining*, MIT Press, Cambridge, Massachusetts.

Kroopnick, P. *et al.*, 1975. *Magnochem 1975, Preliminary Cruise Report Mn-75-01, R/V Kana Keori, Honolulu to Honolulu, 14–25 June 1975*, International Decade of Ocean Exploration, 16 pp. (IDOE Manganese Nodule Project Report No. 11).

LaMotte, R., 1970. Deepsea Ventures pilot run is successful. *Ocean Industry*, October.

Lampietti, F. J. and L. F. Marcus, 1979. Manganese nodules on the sea floor: are economic operations feasible? *Science*, **203**, p. 565.

Lane, A. L., 1977. NOAA's program in deep seabed mining. *Marine Technology Society Journal*, **11**, No. 1.

Lane, A. L., 1979. The environmental aspects of deep ocean mining, in Flipse, J. E., ed., *Deep Ocean Mining, op. cit.*, pp. 33–45.

Lenoble, J. P., 1981. Polymetallic nodules resources and reserves in the north Pacific from the data collected by AFERNOD. *Ocean Management*, pp. 9–24.

Lenoble, J.-P. and P. Rogel, 1977. French activities in the field of polymetallic nodules. *Annales des Mines*, May, pp. 73–80.

Li, T. M. and C. R. Tinsley, 1975. Deep ocean floor nodule mining — first generation techniques are here. *Mining Engineering*, April.

Maline, T. C., C. Garside and D. S. Roels, 1973. Potential environmental impact of manganese-nodule mining in the deep sea. Offshore Technology Conference, *Preprints, OTC*, Dallas, Texas, vol. I, pp. 129–135.

McLeod, C. R. and D. W. Pasho, eds., 1976. Papers presented at the Seventh Underwater Mining Institute, Madison, Wisconsin, 28–29 October, unpublished manuscript.

Marjoram, T. *et al.*, 1981. Manganese nodules and marine technology. *Resources Policy*, **7**, no. 1.

Meiser, H.-J., 1975. International consortia — how to realize ocean mining, in Metallgesellschaft AG, *Review of the Activities, op. cit.*, pp. 71–77.

Melo, U. de and W. Guazelli, 1978. *Manganese Nodules — Importance and Trends*, Petrobas, Rio de Janeiro, Brazil (Projecto Remac), 54 pp. (Portuguese).

Menard, H. W., 1964. *Marine Geology of the Pacific*, McGraw Hill, New York.

Mero, J. L., 1965. *The Mineral Resources of the Sea*, Elsevier Publishing Co., Amsterdam, Holland.

Mero, J. L., 1967. Alternatives for mineral exploitation, in Law of the Sea Institute, *Proceedings of the Second Annual Conference*, University of Rhode Island, Kingston, Rhode Island, pp. 94–97.

Metallgesellschaft AG, 1975. *Review of the Activities: Edition 18 — 1975*,

Manganese Nodules — Metals from the Sea Metallgesellschaft, Frankfurt am Main, Federal Republic of Germany.

Mielke, J. E., 1976. *Ocean Manganese Nodules, 2nd Edition*, Report prepared for U.S. Senate Committee on Interior and Insular Affairs, U.S. Govt. Printing Office, Washington, D.C., 163 pp.

Mills, E. L., ed., 1975. *One Hundred Years of Oceanography*, Dalhousie University, Halifax, Nova Scotia.

Mizuno, A., 1981a. *Deep sea mineral resources investigation in the northern part of central Pacific basin*, Geological Survey of Japan Cruise Report, No. 15, p. 309.

Mizuno, A., 1981b. *Regional and local variabilities of manganese nodules in the central Pacific basin*, Geological Survey of Japan Cruise Report, No. 15, pp. 281–296.

Mizuno, A. and Moritani, T., 1976. Manganese nodule deposits of the central Pacific basin. *World Mining and Metals Technology*, pp. 267–281.

Mizuno, A. and Moritani, T., eds., 1976. *Proceeding of Joint Meeting of the Mining and Metallurgical Institute of Japan and the American Institute of Mining Metallurgical and Petroleum Engineers*, Denver, Colorado.

Moore, J. R., R. P. Meyer and C. L. Morgan, 1973. *Investigation of Sediments and Potential Manganese Nodule Resource of Green Bay Wisconsin*, Sea Grant College Program, University of Wisconsin (Report No. WIS-SG-73-218).

Moore, T. C. and G. R. Heath, 1966. Manganese nodules, topography and thickness of Quaternary sediments in the central Pacific. *Nature*, **212**.

Morgan, M. J., 1978. *Dynamic Positioning of Offshore Vessels*, The Petroleum Publishing Co., Tulsa, Oklahoma.

Morgenstein, M., ed., 1973. *Papers on the Origin and Distribution of Manganese Nodules in the Pacific and Prospects for Exploration*, International Symposium, Honolulu, Hawaii, 23–25 July. Organized by the Valdivia Manganese Exploration Group and the Hawaii Institute of Geophysics.

Moritani, T., 1979. *Deep Sea Mineral Resources Investigation in the Central-Western Part of Central Pacific Basin*, Geological Survey of Japan Cruise Report No. 12, 255 pp.

Moritani, T. and S. Nakao, 1978. *Deepsea Mineral Resources Investigation in the Western Part of Central Pacific Basin January–March 1978.* Geological Survey of Japan Cruise Report No. 17, 281 pp.

Mosnier, J., 1976. Manganese nodules: possible use of geophysical methods. *Oceanis*, 3, No. 8, pp. 418–426 (French).

National Academy of Engineering, Marine Board, 1976. *Seafloor Engineering: National Needs and Research Requirements*, National Research Council, National Academy of Sciences, Washington, D.C., 81 pp.

National Materials Advisory Board, United States Department of

Commerce, 1981. *Manganese Reserves and Resources of the World and Their Industrial Implications: Chapter 4 — Deep Seabed Nodules*, National Academy Press, 360pp. (Publication No. NMAB-374).

National Oceanic and Atmospheric Administration, U.S. Department of Commerce, 1976. *Proceedings of the Marine Minerals Workshop, Silver Spring, Maryland, 23–25 March*, NOAA, Rockville, Maryland.

Nyhart, J. D., *et al.*, 1981. *Toward Deep Ocean Mining in the Nineties, A Description of the Preproduction and Commercial Stages of a Hypothetical Pioneer Venture*, Draft Report for Dept. of Ocean Engineering and Sloan School of Management, Massachusetts Institute of Technology, 1 April 1981. 322 pp.

Ocean Association of Japan, 1979. *The Deep Seabed and Its Mineral Resources: Proceedings of the 3rd International Ocean Symposium, 15–17 November 1978, Tokyo, Japan*, Ocean Association of Japan, Tokyo, Japan.

Ocean Management, Inc., 1982. *Application for and Notice of Claim to Exclusive Exploration Rights for Manganese Nodule Deposits in the Northeast Equatorial Pacific Ocean* (Filed with National Oceanic and Atmospheric Administration, U.S. Department of Commerce, 19 February.

Ocean Minerals Company, 1982. *Ocean Minerals Company Deep Seabed Mining Exploration License Applications* (Filed with National Oceanic and Atmospheric Administration, U.S. Department of Commerce, 25 January).

Organizing Committee, International Ocean Development Conference, 1975. *Third International Ocean Development Conference*, Tokyo, Japan, 5 August.

Pasho, D. W., 1979. Determining deep sea-bed mine-site area requirements—a discussion. United Nations Ocean Economics and Technology Office, *Manganese Nodules: Dimensions and Perspectives, op. cit.*, pp. 83–112.

Pasho, D. W. and D. R. Kerluke, 1977. Striking it rich below. *Canadian Shipping and Marine Engineering*, **48**, no. 4, pp. 35–37.

Paton, A. *et al.*, eds., 1977. *Sea Floor Development: Moving into Deeper Water*, Royal Society of London, U.K.,1978,189 pp. (Proceedings of sea floor development: moving into deeper water (discussion), London, U.K., 1 June).

Patton, D. J., D. Beckton and D. M. Johnston, eds., 1977. *The Future of the Offshore, Legal Developments and Canadian Business*, Centre for International Business Studies, Dalhousie University, Halifax, Nova Scotia.

Pearson, J. S., 1975. *Ocean Floor Mining*, Noyes Data Corporation, Park Ridge, New Jersey.

Piper, D. F., compiler, 1977. *DOMES: Geology and Geochemistry of DOMES Sites A, B and C, Equatorial North Pacific*, U.S. Geological Survey, 527 pp. (USGS Open-File Report No. 77–778).

Prescott, J. H., 1976. Offshore plant economics bared. *Chemical Engineering*, November 8, pp. 86–88.

Raab, W., 1972. Physical and chemical features of the Pacific deep sea manganese nodules and their implications to the genesis of nodules, in Horn, D. R., ed. *Ferromanganese Deposits on the Sea Floor, op. cit.*, pp. 31–50.

Richards, A. F., ed., 1967. *Marine Geotechnique*, University of Illinois Press, Chicago, Illinois, 327 pp.

Rona, P. A., 1978. Plate tectonics, energy and mineral resources: implication for marine engineering. *Society of Naval Architects and Marine Engineers, Symposium*, vol. 2, pp. 2.1–2.11.

Rothstein, A. J., 1970. Deep ocean nodule mining. *Underwater Science and Technology Journal*, September.

Rothstein, A. J. and R. Kaufman, 1973. The approaching maturity of deep ocean mining — the pace quickens, Offshore Technology Conference, *Preprints*, OTC, Dallas, Texas, vol. 1, pp. 323–344.

Schultze-Westrum, H. H., 1973. The station and cruise pattern of the R/V Valdivia in relation to the variability of manganese nodule occurrences, in Morgenstein, M., ed., *Papers on the Origin and Distribution of Manganese Nodules in the Pacific and Prospects for Exploration, op. cit.*, pp. 145–149.

Science Applications, Inc., 1979a. *Technical and Managerial Aspects of Deep Ocean Mining (DOM) Technology Transfer*, SAI, La Jolla, California, 1979 (Briefing presented at the forum provided by the United Methodist Law of the Sea Office, New York, 24 July).

Science Applications, Inc., 1979b. *Study of Deep Ocean Mining Operational Economics*, SAI, La Jolla, California, 1979 (SAI Report No. SAI057-79-918LJ, Prepared for U.S.B.M.).

Seehafen-Verlag Erik Blumenfeld, 1976. *3rd International Conference and Exhibition for Ocean Engineering and Marine Sciences, Dusseldorf, 15 June 1976, Proceedings*, Seehafen-Verlag, Hamburg, (German).

Smale-Adams, K. B. and G. O. Jackson, 1977. Manganese nodule mining. Paper presented at *Seafloor Development: Moving into Deeper Water*, London, U.K., 1 June.

Speiss, F. N. and J. Greenslate, 1976. *Pleiades Expedition Leg-04 MN 76-01, R/V Melville Preliminary Cruise Report*, International Decade of Ocean Exploration, National Science Foundation, Washington, D.C. 87 pp. (IDOE.NSF Technical Report No. 15).

Sumi, K., 1980. Transfer of deep seabed mining technology, proposal on the establishment of deep seabed mining technology bank. *Bulletin of Yokohama City University*, March 1980, pp. 115–138.

Sumitomo Shoji Kaisha, 1972. *Historical Review of Manganese Nodule Development by the Sumitomo Group.*

Taylor, D. M., 1969. New concepts in offshore production. *Ocean Industry*, February.

Third United Nations Conference on the Law of the Sea, 1974a. *Economic Implications of Sea-Bed Mineral Development in the International Area: Report of the Secretary-General*, United Nations, New York (Document A/CONF.62/25 of 25 May 1974).

Third United Nations Conference on the Law of the Sea, 1974b. *Problems of Acquisition and Transfer of Marine Technology. Report of the Secretary-General*, United Nations, New York (Document A/CONF.62/C.3/L.3 of 27 July 1974).

Third United Nations Conference on the Law of the Sea, 1975. *Description of Some Types of Marine Technology and Possible Methods for Their Transfer*: United Nations, New York (Document A/CONF.62/C.3/L.33).

Tinsley, C. R., 1976. A new picture emerges in deep-ocean mining. *Mining Engineering*, April, pp. 34–37.

Tinsley, C. R., 1977. Nodule miners ready for prototype testing. *Engineering and Mining Journal*, January.

Tinsley, C. R., 1978a. Activities and economics of existing manganese nodule mining consortia. Marine Technology Society, *Ocean '78*, MTS, Washington, D.C., pp. 602–605.

Tinsley, C. R., 1978b. *Capital costs for Individual Manganese Nodule Mining Consortia*, Continental Bank, Chicago, Illinois.

Tinsley, C. R., 1979. Manganese nodule mining industry; a study of expected investment requirements, in United Nations Ocean Economics and Technology Office, *Manganese Nodules: Dimensions and Perspectives, op. cit.*, pp. 119–138.

United Nations Division for Economic and Social Information, 1981. *Mining Deep Sea-Bed Minerals*, United Nations, New York (DPI/DESI NOTE/587 of 17 August).

United Nations Ocean Economics and Technology Office, 1979. *Manganese Nodules: Dimensions and Perspectives*, D. Reidel Publishing Co., Dordrecht, Holland.

United States Army Corps of Engineers, 1982. *Dredging — An Annotated Bibliography on Operations, Equipment and Processes*, March, 265 pp. (NTIS Report No. HL-82-7).

United States Department of Commerce, Office of Ocean Minerals and Energy, 1981. *Deep Sea-bed Mining — Marine Environmental Research Plan 1981–1985*.

Victory, J. J., 1976. Mining vessel to be tested in 18,000 ft water in 1977. *Ocean Industry*, August.

Völger, K., 1975. Remote sensing, ecology and marine technology, in Metallgesellschaft AG., *Review of the Activities, op. cit.*, pp. 54–64.

Vugts, J. H., 1970. *The Hydrodynamic Forces and Ship Motions in Waves*, Shipbuilding Department, Delft University of Technology, Delft, Holland (Ph.D. Thesis).

Welling, C. G., 1972. Some environmental factors associated with deep ocean mining. *Eighth Annual Conference of the Marine Technology Society.*

Welling, C. G., 1977. Status of the deep ocean mining program at Lockheed Missiles and Space Co. *Oceans '77* Conference.

Welling, C. G., 1979a. Manganese nodule mining — a risk management assessment, in Flipse, J. E., ed., *Deep Ocean Mining, op. cit.*, pp. 61–73.

Welling, C. G., 1979b. The future outlook for the nodule industry, in United Nations Ocean Economics and Technology Office, *Manganese Nodules: Dimensions and Perspectives, op. cit.*, pp. 139–146.

Welling, C. G., 1980. The ocean's waiting mineral resources. *Stockton's Port Soundings*, August.

Wenzel, J. C., 1968. Systems-development planning, in *Ocean Engineering*, John Wiley and Sons, New York.

Wilmot, P. and A. Slingerland, eds., 1975. *Technology Assessment and the Oceans*, International Conference on Technology Assessment, Monaco.

Willums, J.-O., 1978. Ocean mining: its promises and its problems. *Northern Offshore*, **7**, no. 7, pp. 16–18.

Yamaguchi, U., 1979. Comments from a mining engineer, in Ocean Association of Japan, *The Deep Seabed and Its Mineral Resources, op. cit.*, pp. 113–115.

Yasui, J., 1979. Comments on the success of manganese nodule mining test, in Ocean Association of Japan, *The Deep Seabed and Its Mineral Resources, op. cit.*, pp. 93–94.

B. EXPLORATION AND MINING TECHNOLOGY

Abich, G., 1976. Mooring of a navigation buoy (NAREF) for the deepsea manganese nodules exploration off Hawaii, in *Seehafen-Verlag, op. cit.*, vol. 2, pp. 1023–1033. (German).

Alger, G. R. and D. B. Simons, 1968. Fall velocity of irregular shaped particles. *Journal of the Hydraulics Division, Proceedings of the American Society of Civil Engineers*, p. 721.

Anon., 1972. West Germany uses new TV system for nodule mapping. *Undersea Technology*, **13**, no. 9, p. 43.

Anon., 1975. Prospecting and exploration techniques for ocean resource development, *Mining Engineering*, **27**, no. 4, pp. 36–41.

Anon., 1976. Glomer Explorer's many technical innovations. *Ocean Industry*, **11**, no. 12, pp. 67–73.

Anon., 1977a. Drilling technology applicable to ocean mining. *Ocean Resources Engineering*, September, pp. 8–11.

Anon., 1977b. How photography maps the sea floor to find mineral deposits. *World Mining*, June, pp. 52–53.

Anon., 1977c. Mining for manganese nodules: successful collector proving trials. *Meerestechnik.* **8**, no. 1, pp. 18–19.

Anon., 1981. The underwater bucket dredge for offshore mining. *The Ocean Mining Report,* **14**, no. 11, pp. 12–13.

Anon., 1982. Multibeam bathymetric swath survey systems. *Sea Technology,* June 1982, pp. 28–31.

Assembly of Engineering, Panel on Marine Minerals Technology of the Marine Board, 1977. *Priorities for Research in Marine Mining Technology,* National Research Council, National Academy of Sciences, Washington, D.C., 72pp.

Ball, J., 1967. New concept for lifting nodules. *Ocean Industry,* June.

Bastien-Thiry, H., 1974. Sampling and surveying techniques, in United Nations Ocean Economics and Technology Office, *Manganese Nodules: Dimension and Perspectives, op. cit.,* pp. 7–19.

Beckman, H. and G. Demiray, 1974. The Nanolog, a new way to log the resistivity of the sea floor. *Geophysical Prospecting,* **24**.

Bouma, A. H. and N. F. Marshall, 1964. A method for obtaining and analyzing undisturbed oceanic sediment samples. *Marine Geology,* **2**.

Brink, A. W. and J. S. Chung, 1981. Automatic position control of a 30,000 tons ship during ocean mining operations, Offshore Technology Conference, *Proceedings,* OTC, Dallas, Texas, vol. 3, pp. 205–224.

Brocket, F. H. and W. M. Kollwentz, 1977. An international project —nodule collectors, Offshore Technology Conference, *Proceedings,* OTC, Dallas, Texas, pp. 413–418.

Brocket, F. H., R. A. Petters and G. M. White, 1979. Designing an ocean mining collector system, Offshore Technology Conference, *1979 Proceedings,* OTC, Dallas, Texas, vol. 1, pp. 95–99.

Brockett, F. H. and R. A. Petters, 1980. A proposal for a commercial scale manganese nodule collector design, Marine Technology Society, *Marine Technology '80,* MTS, Washington, D.C., pp. 308–312.

Burns, J. A. and S. L. Suh, 1979. Design and analysis of hydraulic lift systems for deep ocean mining, Offshore Technology Conference, *Proceedings,* OTC, Dallas, Texas, vol. 1, pp. 73–84.

CCOP/SOPAC, Secretariat, 1975. Recent technological progress towards exploitation of minerals from and below the seabed and ocean floor, report for the *Preparatory Meeting for the Establishment of CCOP/ SOPAC* and Proceedings of First and Second Sessions, pp. 66–70.

Chapman, R. R. and P. R. Davies, 1976. Sub-sea surveys, Offshore Technology Conference, *Proceedings,* OTC, Dallas, Texas.

Chaziteodorou, G., W. Strangler and A. Wienen, 1975. The hydro-pneumatic haulage of manganese nodules from the deep-sea. *Meerestechnik,* April, pp. 60–65.

Chaziteodorou, G., W. Strangler and A. Wienen, 1977. The hydro-pneumatic haulage of manganese nodules from the deep-sea. *Meerestechnik,* February, pp. 9–18.

Chung, J. S., 1976. Hydrodynamic forces on a marine riser: a velocity potential method, Presented at the *Joint Petroleum Mechanical Engineering and Pressure Vessels and Piping Conference*, Mexico City, Mexico, September (American Society of Mining Engineers (ASME) Paper No. 76-Pet-81).

Chung, J. S., 1981a. Nonlinear static analysis of deep ocean mining pipe. *Journal of Energy Resources Technology*, March (American Society of Mining Engineers (ASME) Transactions).

Chung, J. S., 1981b. Nonlinear static analysis of deep ocean mining pipe: part II. numerical study. *Journal of Energy Resources Technology*, March (American Society of Mining Engineers (ASME) Transactions).

Chung, J. S. and A. K. Whitney, 1980. Nonlinear transient motion of deep-ocean mining pipe, Offshore Technology Conference, *1980 Proceedings*, OTC, Dallas, Texas.

Chung, J. S. and A. K. Whitney, 1981. Dynamic vertical stretching oscillation of an 18,000-ft. ocean mining pipe, Offshore Technology Conference, *Proceedings*, OTC, Dallas, Texas.

Clauss, G., 1972. Some investigations into airlift systems for mineral recovery in ocean mining. *Erdöl-Erdgas*, August.

Clauss, G., 1975. Slip and friction losses in deep sea hydraulic lifting of solids, in Organizing Committee, International Ocean Development Conference, *Third International Ocean Development Conference Proceedings, op. cit.*

Clauss, G., 1978. Hydraulic lifting in deep-sea mining. *Marine Mining*, 1, no. 3, pp. 189–208.

Cruickshank, M. J., 1975. Technological and Environmental Considerations in the Exploration and Exploitation of Marine Minerals, University of Wisconsin, Madison, Wisconsin (Unpublished Ph.D. Thesis).

Dane, E. B., 1973. Harvesting manganese nodules; U.S. patent 3675348 (11th July 1972). *Underwater Journal and Information Bulletin*, 5, No. 1, p. 39.

Dedegil, M. Y., 1976. The jet lift system in deep sea mining, in *Seehafen-Verlag, op. cit.*, vol. 1, pp. 146–153. (German).

Defossez, M., *et al.*, 1976. Contribution of statistical methods to the study of some South Pacific manganese nodule deposits and associated sediments (abstract). *Abstracts of 25th International Geological Congress*, vol. 2, pp. 342–343.

Denning, R. A., 1965. The flow of solids-water mixtures in hydraulic dredging. *Symposium on Rheology*, June.

Dennis, A. R., 1975. Satellite positioning and navigation for off-shore applications: past, present and future, Offshore Technology Conference, *Proceedings*, OTC, Dallas, Texas, pp. 243–253.

Dettweiler, T. K. and G. A. Zahn, 1980. Navigation and positioning for an ocean mining system, Marine Technology Society, *Marine Technology '80*, MTS, Washington, D.C., pp. 358–361.

Earl and Wright, 1972. Harvesting manganese nodules, U.S. patent 3672725 (27th June 1972), *Underwater Journal and Information Bulletin*, 5, no. 1, p. 43.

Engelmann, H. E., 1978. Vertical hydraulic lifting of large-size particles — a contribution to marine mining, Offshore Technology Conference, *1978 Proceedings*, OTC, Dallas, Texas, vol. 2, pp. 731–740.

Ensign, C. O., 1969. Operational aspects of ocean mining. *Mining Congress Journal*, 55, no. 2, pp. 93–98.

Ewing, M., et al., 1971. Photographing manganese nodules on the ocean floor. *Oceanology International*, 6, no. 12.

Fagot, M. G., et al., 1981. Deep-towed seismic system design for operation at depths up to 6000 m, Offshore Technology Conference, *1981 Proceedings*, OTC, Dallas, Texas, vol. 3, pp. 141–154.

Fellerer, R., 1975. Methods and problems of the exploration of manganese nodules. *Oceanology International 75*.

Flanagan, F. J. and D. Gottfried, 1980. *USGS Rock Standards III: Manganese Nodule Reference Samples USGS-Nod-A-1 and USGS-Nod-P-1*, U.S. Government Printing Office, Washington, D.C. (USGS Professional Paper 1155).

Fleischmann, W. and M. von Heimendahl, 1977. Electron microscopic investigation on a Pacific manganese concretion. *Mineralogische Deposita*, 12, no. 2, pp. 155–162.

Flipse, J. E., 1969. An engineering approach to ocean mining, Offshore Technology Conference, *Preprints*, OTC, Dallas, Texas, vol. I, pp. 317–332.

Flipse, J. E., 1970. Development in ocean exploration and mining. *Mining Congress Journal*.

Friedrich, G. H. W., H. Kunzendorf and W. L. Pluger, 1974. Shipborne geochemical investigation of deep-sea manganese nodule deposits in the Pacific using a radioisotope energy-dispersive x-ray system. *Journal of Geochemical Exploration*, 3, pp. 303–317.

Gauthier, M., 1977. Mining on the continuous line bucket (CLB) system, in European Economic Community, Paper presented at Seminar on the Exploitation of the Deep Seabed, *op. cit.*

Gauthier, M., 1978. The exploration of the manganese nodules. *Revue Maritime*, no. 340, pp. 2353–2369. (French).

Geminder, R. and E. J. Lecourt, 1972. Deep ocean mining system tested. *World Dredging and Marine Construction*, 8, no. 8, pp. 35–38.

Golan, L. P. and A. H. Stenning, 1969. Two-phase vertical flow maps. *Proceedings of Institute of Mechanical Engineers*, 184, (Paper no. 14).

Goring, H. O., 1976. The submersible motor and its application in ocean engineering, in *Seehafen-Verlag, op. cit.*, vol. 1, pp. 721–733.

Grote, P. B. and J. A. Burns, 1981. System design considerations in deep ocean mining lift systems. *Marine Mining*, 2, no. 4, pp. 357–383.

Grubbs, K. W., 1981. *Sound Scattering Measurement from a Single Manganese Nodule*, Aerospace and Ocean Engineering Dept.,

Virginia Polytechnic Institute and State University, Blacksburg, Virginia (M.S. Thesis).

Hagerty, R. Usefulness of spade corer for geotechnical studies and some results from the northeast Pacific, in Inderbitzen, A. L., ed., *Deep Sea Sediments*, Plenum Press, New York, pp. 169–186.

Halkyard, J. E., 1980. Ore handling and transfer at sea, Marine Technology Society, *Marine Technology 80*, MTS, Washington, D.C., pp. 303–307.

Handa, K., Y. Miyashita and S. Oba, 1982. Design and observation tests of an experimental nodule collection vehicle, Offshore Technology Conference, *1982 Proceedings*, OTC, Dallas, Texas, vol. 2, pp. 457–464.

Harris, D., 1982. Underwater mining methods and applications, *World Dredging and Marine Construction*, July, pp. 30–32.

Hasegawa, K., 1979, History and outlook of the development of production technology of manganese nodules, in Ocean Association of Japan, *The Deep Seabed and Its Mineral Resources, op. cit.*, pp. 79–83.

Heine, O. R. and S. L. Suh, 1978. An experimental nodule collection vehicle design, Offshore Technology Conference, *1978 Proceedings*, OTC, Dallas, Texas, vol. 2, pp. 741–750.

Herbich, J. B. and J. E. Flipse, 1978. Technological gaps in deep ocean mining, Marine Technology Society, *Oceans '78*, MTS, Washington, D.C.

Hering, N., 1973. New knowledge on prospecting and exploration of ore nodule deposits. *Meerestechnik*, 4, no. 1, pp. 1–11.

Hess, G., 1976. Evaluation methods and results of the exploration for manganese nodules, in *Seehafen-Verlag, op. cit.*, vol. 1, pp. 42–58.

Hirota, T., 1979. Present state and problems of technology research and development, in Ocean Association of Japan, *The Deep Seabed and Its Mineral Resources, op. cit.*, pp. 83–88.

Hirst, T. J. and A. F. Richards, 1975. Analysis of deep-sea nodule mining — seafloor interaction, Offshore Technology Conference, *Proceedings*, OTC, Dallas, Texas, pp. 923–932.

Institution of Mining and Metallurgy, 1979. *Developing Technology Relating to Ship's Motion, Behaviour of Bulk Mineral Cargoes and Deep-Sea Mining*, IMM, London, 24 pp.

Isaacs, C. R., 1974. Dredging of bulk samples of manganese nodules. *Mining Engineering*, 6, no. 4, pp. 27–30.

Jacobsen, B. K. and K. Kure, 1979. Deep sea mining model tests, Offshore Technology Conference, *1979 Proceedings*, OTC, Dallas, Texas.

Jacobsen, B. K. and K. Kure, 1981. Computer simulation of deep sea mining manoeuvres, Offshore Technology Conference, *1981 Proceedings*, OTC, Dallas, Texas, vol. 3, pp. 235–246.

Jepsen, J. C. and J. L. Ralph, 1969. Hydrodynamic studies of two-phase upflow in vertical pipelines, *Proceedings of Institute of Mechanical Engineers*, **184** (Paper No. 20).

Johnson, P. A., 1970. The design of ocean mineral prospecting ships, Offshore Technology Conference, *Preprints*, OTC, Dallas, Texas, vol. II, pp. 95–106.

Johnson, P. A., 1971. *Studies of Deep Sea Erosion Using Deep Towed Instrumentation*, Scripps Institution of Oceanography, San Diego, California, (SIO Reference 71-24).

Jones, O. C., Jr. and N. Zuber, 1975. The interrelation between void fraction fluctuations and flow patterns in two-phase flow. *International Journal of Multiphase Flow*, 2, pp. 273–306.

Kato, H. S. T. and T. Miyazawa, 1975. A study of an air-lift pump for solid particles and its application to marine engineering, *Second Symposium on Jet Pumps and Ejectors and Gas Lift Techniques*, 24–26 March (Paper 63).

Kaufman, R., 1971. Deep ocean exploitation technology, *Marine Technology*, April, pp. 145–158.

Kaufman, R. and A. J. Rothstein, 1970. Recent developments in deep ocean mining, *Sixth Annual Conference of the Marine Technology Society Preprints*, MTS, Washington, D.C.

Kaufman, R. and J. P. Latimer, 1971. The design and operation of a prototype deep-ocean mining ship. *Transactions of the Society of Naval Architects and Marine Engineers*, 70, p. 4.

Kaufman, R. and W. D. Siapno, 1972. Future needs of deep ocean mineral exploration and surveying, Offshore Technology Conference, *Preprints*, OTC, Dallas, Texas, pp. 309–332.

Kollwentz, W., 1975. Prospecting and exploration of manganese nodule occurrences, in Metallgesellschaft AG, *Review of the Activities, op. cit.*, pp. 12–26.

Kollwentz, W., 1979. Development of deep ocean mining system — a multi-national project in Ocean Association of Japan, *The Deep Seabed and Its Mineral Resources, op. cit.*, pp. 74–79.

Kroonenberg, H. H. van den, 1979. A novel vertical underwater lifting system for manganese nodules using a capsule pipeline, Offshore Technology Conference, *1979 Proceedings*, OTC, Dallas, Texas, vol. 1, pp. 59–72.

Krutein, M. G., 1978. Ocean mining and transport technology, Paper presented at the Seminar 'Deep Seabed Mining of Manganese Nodules', McGill University, Montreal, Canada, 13–17 March 1978 (mimeographed).

Krutein, M. G., 1981. Ocean Mining Technology, in Borgese E. M. and P. M. T. White eds, *Seabed Mining: Scientific, Economic, Political Aspects*, International Ocean Institute, Malta (I.O.I. Occasional Papers No. 7).

Kuntz, G., 1979. The technical advantages of submersible motor pumps in deep sea technology and the delivery of manganese nodules, Offshore Technology Conference, *1979 Proceedings*, OTC, Dallas, Texas, vol. 1, pp. 85–94.

Lange, J. and W. G. Biemann, 1976. Development of components for *in-situ* analysis system for the exploration of manganese nodules, Offshore Technology Conference, *Proceedings*, OTC, Dallas, Texas, pp. 585–589.

Lange, J., U. Tamm and H. Wurz, 1975. Development of a towed bottom device for *in-situ* analysis of manganese nodule fields. *Meerestechnik*, April, pp. 50–55.

Leclaire, L., 1976. Manganese nodules: lifting systems from the sea floor, for studying the concentration and content. *Oceanis*, **3**, No. 8, pp. 428–462 (French).

Lecourt, E. J., Jr. and D. W. Williams, 1971. Deep ocean mining — new application for oil field and marine equipment, Offshore Technology Conference, *Preprints*, OTC, Dallas, Texas, vol. I, pp. 859–874.

Lee, H. J., 1974. The role of laboratory testing in the determinations of deep-sea sediment engineering properties, Inderbitzen, A. L., ed., *Deep-Sea Sediments*, pp. 111–127.

Leest, H. van and O. Bussemaker, 1976. The performance and characteristic of thrusters, Paper presented at *Offshore South Asia Conference*, (Paper No. 40).

Lense, A. H., 1968. Sampling minerals of the ocean floor. *Mining Engineering*, **20**, no. 8, pp. 54–57.

Ma, Y., 1982. *Acoustic Scattering Analysis for Remote Sensing of Manganese Nodules*, Aerospace and Ocean Engineering Dept., Virginia Polytechnic Institute and State University, Blacksburg, Virginia (Ph.D. Thesis).

Ma, Y. and A. H. Magnuson, 1981. *Acoustic Scattering from a Single Manganese Nodule*, Aerospace and Ocean Engineering Dept., Virginia Polytechnic Institute and State University (Technical Report).

Magnuson, A. H., *et al.*, 1981. Acoustic sounding for manganese nodules, Offshore Technology Conference, *Proceedings*, OTC, Dallas, Texas, pp. 147–160.

Magnuson, A. H., *et al.*, 1982. Remote acoustic sensing of manganese nodule deposits, Offshore Technology Conference, *1982 Proceedings*, OTC, Dallas, Texas, vol. 2, pp. 431–444.

Marsh, J. B., 1980. *Technological Development in the Manganese Nodule Industry*, Paper presented at the Tenth Atlantic Economic Conference. Boston, Massachusetts, October (unpublished manuscript).

Marsh, J. B., 1981. Technological development in the manganese nodule industry. *Atlantic Economic Journal*, March.

Maruyama, S. and Y. Kinoshita, 1975. Deep sea-bed manganese nodule and a grab-type mud sampler, OKEAN-70. *Chishitsu-Nyusu*, **250**, pp. 1–7. (Japanese).

Masuda, Y., 1972a. Development work to deep sea resources of manganese nodule using continuous line bucket (CLB) system by Japanese group and its future, *Second International Ocean Development Conference*, Tokyo, Japan, 5–7 October.

Masuda, Y., 1972b. Endless chain scoop. U.S. patent 3672079 (27th June 1972). *Underwater Journal and Information Bulletin*, **5**, no. 1, p. 38.

Masuda, Y., M. J. Cruickshank and J. L. Mero, 1971. Continuous bucket line dredging at 12,000 feet, Offshore Technology Conference, *Preprints*, OTC, Dallas, Texas.

Mateker, E., Jr. and L. E. Ott, 1973. Deep water navigation with an inertial system, Offshore Technology Conference, *Proceedings*, OTC, Dallas, Texas.

McFarland, W. D., 1980. *Development of a Reliable Method for Resource Evaluation of Deep-Sea Manganese Nodule Deposits Using Bottom Photographs*, Department of Geology, Washington State University, (M.S. Thesis).

Mero, J. L., 1971. Recent concepts in undersea mining, *American Mining Congress*, Las Vegas, Nevada, 4 August.

Mizuno, A., J. Chujo and N. Yamakado, 1975. Prospecting of manganese nodule deposits of the eastern central Pacific basin, in Organizing Committee, International Ocean Development Conference, *op. cit.*, vol. 3, pp. 321–334.

Moore, J. R. and M. J. Cruickshank, 1973. *Identification of technological gaps in exploration of marine ferromanganese deposits*, National Science Foundation (Sea-bed Assessment Program, Inter-university program of research on ferro-manganese deposits on the ocean floor), (Mimeograph manuscript).

Morel, Y. G., 1975. Dredging of manganese nodules by the CLB system, in Report for the *Preparatory Meeting for the Establishment of CCOP/SOPAC* and Proceedings of First and Second Sessions, pp. 134–136.

Moritani, T. and F. Murakami, 1979. Relation between manganese nodule abundance and acoustic stratigraphy in the GH 77-1 area in Geological Survey of Japan, *Cruise Report No. 12*, pp. 218–221.

Mudie, J. D., J. A. Grow and J. S. Bessey, 1972. A near-bottom survey of lineated abyssal hills in the equatorial Pacific. *Marine Geophysical Researches*, pp. 397–411.

Nassos, G. P. and S. G. Bankoff, 1967. Slip velocity ratios in an air–water system under steady-state and transient conditions. *Chemical Engineering Science*, **22**, pp. 661–668.

National Oceanic and Atmospheric Administration, 1980. *Workshop on Ocean Acoustic Remote Sensing*, NOAA, U.S. Department of Commerce, Rockville, Maryland.

Neil, S. T., 1980. United States Geological Survey, chemical analysis of USGS manganese nodule reference samples. *Geostandards Newsletter*, **4**, no. 2, pp. 205–212.

Noorany, I., 1972. Underwater soil sampling and testing, state-of-the-art review, ASTM Special Technical Publication (No. SPT501).

Noorany, I. and J. T. Fuller, 1982. Soil mechanic interaction studies for manganese nodule mining, Offshore Technology Conference, *1982 Proceedings*, OTC, Dallas, Texas, vol. 2, pp. 445–456.

O'Brien, M. D., B. A. Kriegsman and E. St. George, 1978. Navigation systems for deep ocean mining, Offshore Technology Conference, *Proceedings*, OTC, Dallas, Texas, pp. 441–448.

Ochi, M. K. and S. L. Bales, 1977. Effects of various spectral formulations in predicting responses of marine vehicles and ocean structures, Offshore Technology Conference, *Proceedings*, OTC, Dallas, Texas.

Oedjoe, D. and R. H. Buchanan, 1966. The pressure drop in hydraulic lifting of dense slurries of large solids with wide size distribution. *Transactions of Institute of Chemical Engineers*, **44**, (Paper T364).

Orkiszewski, J., 1967. Predicting two-phase pressure drops in vertical pipe. *Journal of Petroleum Technology*, June.

Ott, L. E., 1976. A totally integrated navigation system, Offshore Technology Conference, *Proceedings*, OTC, Dallas, Texas, p. 414.

Ozturgut, E., *et al.*, 1978. *Deep Ocean Mining of Manganese Nodules in the North Pacific: Pre-mining Environmental Conditions and Anticipated Mining Effects*, National Oceanic and Atmospheric Administration, U.S. Department of Commerce, 138 pp. (NOAA Technical memorandum, Marine Ecosystem Analysis Program).

Paluzzi, P. R., *et al.*, 1976. Computer image processing in marine resource exploration, Marine Technology Society-Institute of Electrical and Electronic Engineering, *Oceans '76*, pp. 4D-1–4D-10.

Pasho, D. W., 1979. *A Qualitative Consideration of Some Mining Machine Seafloor Interactions*, Dept. of Energy Mines and Resources, Govt. of Canada, Ottawa.

Patterson, R. B., 1972. Inspection of manganese deposits in deep water in Horn, D. R., ed., *Ferromanganese Deposits in the Sea-Floor, op. cit.*, pp. 251–262.

Pepelnik, R., U. Fanger and W. Michaelis, 1976. Nuclear measuring techniques in mining and processing of manganese nodules, Marine Technology Society-Institute of Electrical and Electronic Engineering, *Oceans '76*, pp. 2C-1–2C-6.

Peterson, D. R. and P. B. Grote, 1979. Deep ocean mining system operational simulation, in Flipse, J. E., ed., *Deep Ocean Mining, op. cit.*, pp. 1–21.

Petrick, M. and A. A. Kudrika, 1966. On the relationship between the phase distributions and relative velocities in two-phase flow, *Third International Heat Transfer Conference Proceedings*, Chicago, Illinois.

Petters, R. A. and F. H. Brockett, 1980. Factors affecting the design and selection of a commercial nodule collector, Paper presented at the *Energy Technology Conference and Exhibition*, New Orleans, Louisiana, February 1980.

Pfeiffer, D., ed., 1980. Die Fahrten des Forschungsschiffes Valdivia 1971–1978. *Geowissenschaftliche Ergebnisse* (German).

Richards, A. F., T. J. Hirst and J. M. Parks, 1976. Preliminary geotechnical properties, northeast central Pacific nodule mining area, Marine Technology Society, *Oceans '76*, MTS, Washington, D.C.

Richards, A. F. and J. M. Parks, 1977. Geotechnical predictor equations for east central north Pacific nodule mining area sediments, Offshore Technology Conference, *Proceedings*, OTC, Dallas, Texas, pp. 377–386.

Riedel, O., 1975. Mining and transport of manganese nodules — technological problems in Metallgesellschaft AG., *Review of the Activities, op. cit.*, pp. 36–43.

Robb, J. M., R. E. Sylvester and R. Penton, 1981. Simplified method of deep-tow seismic profiling. *Geo-Marine Letters*, 1, pp. 65–67.

Robinson, C. W., 1971. Supertanker loads ore slurry at sea. *Ocean Industry*, August.

Rose, H. E. and R. A. Duckworth, 1969. Transport of solid particles in liquids and gases. *The Engineer*, 14, 21, 28 March.

Roseman, D. P., A. K. Anderson and J. Lisnyk, 1978. Marine transport of bulk commodities in slurry form — an assessment, *SNAME-STAR Symposium*, Paper No. 8.

Rossfelder, A. M. and N. F. Marshall, 1967. Obtaining large undisturbed and oriented samples in deep water, in Richards, A. F., ed., *Marine Geotechnique, op. cit.*, pp. 243–263.

Rothman, S. and R. J. Lynch, 1978. Utilization of available technology to support ocean mining, Marine Technology Society, *Ocean '78*, MTS, Washington, D.C.

Rothstein, A. J., 1970. Deep ocean nodule mining. *Underwater Journal*, 2, no. 3.

Sato, M., M. Igarashi and M. Hanada, 1977. Performance test of CLB system collecting manganese nodules from seabed. *Journal of Faculty of Marine Science and Technology, Tokai University*, pp. 243–254.

Schatz, C. E., 1971. Observations of sampling and occurrence of manganese nodules, in Offshore Technology Conference, *Preprints*, OTC, Dallas, Texas, vol. I, pp. 389–396.

Schimmelbusch, H. and P. Danzer, 1975. The use of computers in evaluating the results of manganese nodule exploration, in Metallgesellschaft AG, *Review of the Activities, op. cit.*, pp. 50–53.

Sender, F., 1973. Optimal utilization of groundbase navigation systems in world-wide deep sea navigation, Offshore Technology Conference, *Proceedings*, OTC, Dallas, Texas.

Sender, F., 1976. Hydrospheric navigation and positioning in survey missions where Navaids are lean and far off, Offshore Technology Conference, *Proceedings*, OTC, Dallas, Texas, pp. 229–240.

Sender, F., 1977. The NAREF buoy, a deep-sea navigation aid, Offshore Technology Conference, *Proceedings*, OTC, Dallas, Texas, pp. 39–48.

Sharpe, R. T. and P. G. Galyean, 1977. An acoustic navigation system for site survey and geodetic positioning, Offshore Technology Conference, *Proceedings*, OTC, Dallas, Texas, pp. 53–60.

Siapno, W. D., 1975. Exploration technology and ocean mining parameters, *American Mining Congress Convention*, San Francisco, California, 28 September–1 October.

Smith, K., 1981. *Properties of Oceanic Manganese Nodule Fields Relevant to a Remote Acoustical Sensing System*, Aerospace and Ocean Engineering Dept., Virginia Polytechnic Institute and State University, Blacksburg, Virginia (M.S. Thesis).

Smith, K. and K. Sumdkvist, 1980. *Wavespread of Manganese Nodule Material and Statistical Properties of Nodule Deposits*, Aerospace and Ocean Engineering Dept., Virginia Polytechnic Institute and State University, Blacksburg, Virginia, (mimeographed manuscript).

Sorensen, P. E. and W. J. Mead, 1970. Evaluation of technological spillovers—the case of the deep sea dredge, in Marine Technology Society, *Marine Technology 1970*, vol. 2, pp. 779–788.

Spiess, F. N. and R. C. Tyce, 1973. *Marine Physical Laboratory Deep Tow Instrumentation System*, Scripps Institution of Oceanography, San Diego, California, 37 pp. (SIO Reference 73-4, MPL No. U-69/72).

Spiess, F. N., C. D. Lowenstein and D. E. Boegemen, 1978. Fine grained deep ocean survey techniques, Offshore Technology Conference, *Proceedings*, OTC, Dallas, Texas, pp. 715–724.

Spradley, L. H., 1976. Analysis of position accuracies from satellite systems — a 1976 update, Offshore Technology Conference, *Proceedings*, OTC, Dallas, Texas, pp. 425–429.

Sternberger, W. I. and L. R. LeBlanc, 1976. Short range precision navigation and tracking system, Marine Technology Society — Institute of Electrical and Electronic Engineering, *Oceans '76*, pp. 5E1–5E5.

Thijssen, T., *et al.*, 1981. Reconnaissance survey of manganese nodules from the northern sector of the Peru Basin. *Marine Mining*, **2**, no. 4, pp. 385–427.

Tisot, J. P., 1981. Analysis of physical and mechanical properties of deep sea sediments from potential manganese nodule mining areas in the north central Pacific, Offshore Technology Conference, *1981 Proceedings*, OTC, Dallas, Texas, vol. 4, pp. 139–146.

Vigil, A. E., H. Frisbee and G. L. Hatchett, 1975. Deep sea survey system, Offshore Technology Conference, *Proceedings*, OTC, Dallas, Texas.

Wallis, G. B., 1969. *One-Dimensional Two-Phase Flow*, McGraw Hill, New York.

Wang, F. F. H. and M. J. Cruickshank, 1969. Technologic gaps in exploration and exploitation of sub-sea mineral resources, Offshore Technology Conference, *Preprints*, OTC, Dallas, Texas, vol. I, pp. 285–298.

Weber, M. and Y. Dedegil, 1976. Transport of solids according to the air-lift principle, *Fourth International Conference on the Hydraulic Transport of Solids in Pipes*, 18–21 May.

Weiss, R. F., O. H. Kirstein and R. Ackermann, 1977. Free vehicle

instrumentation for the *in situ* measurement of processes controlling
the formation of deep-sea ferromanganese nodules, Paper presented
at *Annual Combined Conference of MTS/IEEE*, Los Angeles,
California, 17 October.

Welling, C. G., 1980a. The massive technology for ocean mining.
Stockton's Port Soundings, August 1980, pp. 8–11.

Welling, C. G., 1980b. The ships and ports. *Stockton's Port Soundings*,
August 1980, pp. 7–9.

Welling, C. G., 1981. An advanced deep sea mining system, Offshore
Technology Conference, *1981 Proceedings*, OTC, Dallas, Texas, vol. 3,
pp. 247–255.

Whitney, A. K., J. S. Chung and B. K. Yu, 1981. Vibration of a long
marine pipe due to vortex shedding. *Journal of Energy Resources
Technology*.

Williams, D. W., C. M. McBride and S. C. Kinnaman, 1977. Deep ocean
mining technology transfer from and to offshore drilling industry,
Offshore Technology Conference, *Proceedings*, OTC, Dallas, Texas,
pp. 395–400.

Wilson, K. C., N. P. Brown and M. Streat, 1979. Hydraulic hoisting at
high concentration: a new study of friction mechanisms. *Sixth
International Conference on the Hydraulic Transport of Solids in
Pipes*, 26–28 September.

Wogman, N. A., H. G. Rieck and H. L. Nielson, 1973. *In situ* analysis of
the major and minor elements in manganese fields. *Marine
Technology Society Journal*, 7, no. 6, pp. 35–40.

Yamakado, N., 1979. Continuous line bucket (CLB) mining system, in
Ocean Association of Japan, *The Deep Seabed and Its Mineral
Resources, op. cit.*, pp. 90–92.

Yamakado, N., K. Handa and T. Usami, 1978. Model tests on continuous
line bucket mining system, Offshore Technology Conference, *1978
Proceedings*, OTC, Dallas, Texas, vol. 2, pp. 725–730.

Zandi, I. and G. Gavatos, 1976. Heterogeneous flow of solids in pipe lines.
*Journal of Hydraulic Division, Proceedings of the American Society of
Civil Engineers*.

Zuber, N. and J. A. Findlay, 1965. Average volumetric concentration in
two-phase flow systems. *Journal of Heat Transfer, Transactions of the
American Society of Mining Engineers (ASME)*.

Zuber, N., *et al.*, 1967. *Steady State and Transient Void Fraction in
Two-Phase Flow Systems*, U.S. Atomic Energy Commission (AEC)
Contract AT 04-3).

Notes and References

1. The deposition of the mineral also occurs as a crust, in the form of either a layer coating an object on the ocean floor or a pavement covering a portion of the ocean floor. So far, the greatest interest has been in the nodules; this volume, thus, deals with the nodules.
2. The terms "ocean floor", "seabed", "seafloor", "deep seabed" etc. are used synonymously in this volume. Manganese nodules have been known to occur in shallow water as well as on the floor of lakes. But the deep seabed nodules have the greatest economic potential, by far.
3. Details about the composition and the activities of these entities can be found in United Nations Department of International Economic and Social Affairs, *Sea-Bed Mineral Resource Development: Recent Activities of the International Consortia.* United Nations, New York, 1980 (UN publication Sales No. E.80.II.A.9) and *Sea-Bed Mineral Resource Development*, UN, New York, 1982 (UN publication Sales No. E.80.II.A.9/Add.1). Appendix B, Table B1 of this volume lists the component members of these entities.
4. Technologies for processing the ore will be dealt with in Volume 3 of the Series, *Analysis of Processing Technology for Manganese Nodules.*
5. This volume does not deal with costs of technology or environmental impact of the application of technology. The former will be discussed in Volumes 6 and 7 of the Series. Preliminary investigations regarding environmental impact indicate that the impact is not significant. However, further investigations need to be carried out before any definitive conclusions can be reached.
6. For details, see United Nations Ocean Economics and Technology Office (1979).
7. Krutein (1981 and 1978) have excellent discussions on the environment of deposition and its influence on the design of mining technology.
8. For details, see Tinsley (1978a), Tinsley (1978b), Tinsley (1979) and Nyhart *et al.* (1981).
9. Mining technology is dealt with in detail in Chapter 5 of the volume.
10. See Bailly (1976).
11. Magnetic devices, e.g. magnetometers, are also used. They are helpful in studying the magnetic characteristics of the crust and sub-bottom features.

12. The collector sub-system is composed of the collector itself and related apparatus and instruments. The collector is referred to in various ways: "miner", "bottom miner" etc.

13. The materials in this Appendix have been collected from Oliver S. North, *Mining Exploration, Mining and Processing Patents*, 1970–1980, Society of Mining Engineers of the American Institute of Mining, Metallurgical, and Petroleum Engineers, Inc., New York, various years. The materials are being reproduced with permission from the SME of AIME.

List of Figures

List of Tables

Index